AF335652

BOOKS PUBLISHED BY THE RATEAVERS

Organic Method Primer, by Bargyla and Gylver Rateaver

A condensation of: *Bio-Dynamic Farming and Gardening* by E. E. Pfeiffer, M.D.

Fertility Pastures & Cover Crops, by Newman Turner

Organic Small Farming, by Hugh Corley

Comfrey Report & (bound together with) *Comfrey, the Herbal Healer* by Lawrence D. Hills

Seaweed in Agriculture & Horticulture, by W. A. Stephenson

Gold in the Grass, by Margaret Leatherbarrow

CONSERVATION GARDENING AND FARMING® SERIES
SERIES C: REPRINTS • Edited by Bargyla and Gylver Rateaver

AGRICULTURE
The only right approach
by P.H. HAINSWORTH

Science says there IS a difference

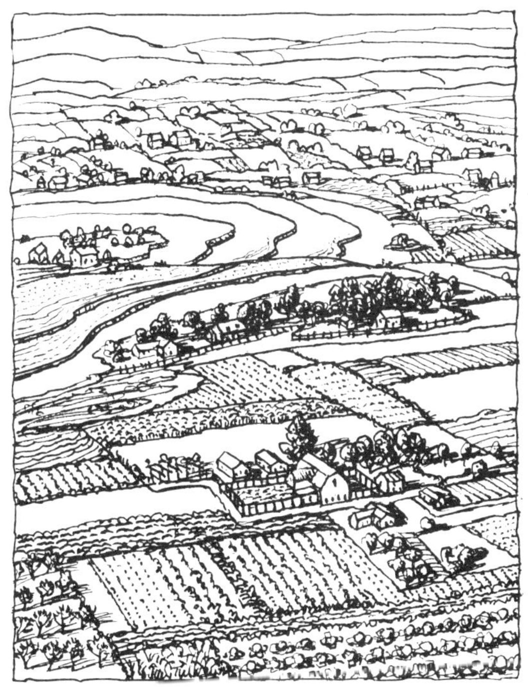

SECOND EDITION - 1976
Bargyla and Gylver Rateaver, Pauma Valley, Calif. 92061

Cover design and title page by Jon Gnagy

FOREWORD

My statement that there are ample "hard" data to support the scientific basis of biological agriculture has been met with stony recalcitrance.

I submit this fact-packed book as an example of such heavily-documented data for the organic method.

No researcher may claim unbiased subservience to science who willfully closes his eyes to these truths.

The original bibliography has been omitted because the listed books and publications are almost all out of print.

The Author, P. H. Hainsworth, 1975

CONTENTS

Introduction	*page*	11
I. The Food of the Plant		13
II. Soils and their Flora and Fauna		36
III. Organic Agriculture		57
IV. Providing the Food		70
V. The Problems of Present Methods of Agriculture		104
VI. Practical Organic Farming		132
VII. Organic Market Gardening		156
VIII. Composting		172
IX. Disease Resistance in Plants		191
X. The Food of the Animal		216
XI. Organic Agriculture and the Future		230
Glossary		237
Bibliography		239
Index		242

ILLUSTRATIONS

1. Soil formation by natural agencies   *facing page* 32
2. Structure of a clover root nodule   33
3. Mycorrhiza in a crop plant   48
4. Phosphate-dissolving organisms   49
5. Soil structure   64
6. Plant roots in the soil   65
7. Effects of worms on soils   80
8. Effects of worms on soils   81
9. (*a*) Effect of soil flora on aggregation of soil crumbs   96
   (*b*) The effect of adding worms to clay subsoil   96
10. Effects of organic matter and cultivation on soils (I)   97
11. Effect of grass roots on soil structure   112
12. Soil fauna   113
13. Market garden crops from a surface-cultivated soil well supplied with organic matter
   (*a*) tomatoes (*b*) strawberries   116
14. Market garden crops from a surface-cultivated soil well supplied with organic matter
   (*a*) parsnips (*b*) lettuce   117
15. Effect of cultivation on soil   124
16. Effects of organic matter and cultivations on soils (II)   125
17. (*a*) Availability of potash in a soil enriched with organic matter   144
17. (*b*) Effect of organic matter on plant growth in poor soils   144
18. Sweet clover   145
19. Sweet clover   160
20. Fruit Trees under organic methods   161
21. Effect of mulching under warm dry conditions   176
22. Soft fruits under organic methods   177
23. Disease resistance under organic methods   192
24. Potato eelworm trial   193

INTRODUCTION

This book is an attempt to gather together relevant material that may have some bearing on the results achieved by the increasing number of agriculturists practising 'organic' methods. These methods have not so far received the attention deserved from research organizations largely because of the inconsistency of the methods used and confusion regarding the principles involved. It is hoped that this book will elucidate some of the basic principles and reasons for their success and suggest possible lines of research for their further clarification.

The book is mainly concerned with the basis of agriculture, the soil, and its relationship with the crops that grow upon it. While it deals primarily with the problems of farming in temperate climates, the principles set forth apply with equal or even greater intensity to other areas of the world, and particularly to the under-developed areas where, for financial reasons, the use of fertilizers and machinery is difficult.

The term 'Organic' has been retained throughout, as, although it has acquired something of a stigma in some quarters, its sense and significance has become generally accepted by common usage. It is felt, however, that there is room for a more informative expression: it may be that 'Biological' agriculture would be more appropriate.

I have to acknowledge assistance from a number of quarters, particularly the following, for information concerning their farming methods: The Viscount Newport, Lt.-Col. Sir J. E. H. Neville, Bart., Lt.-Gen. H. R. S. Massey, M. R. Coward, Mr. I. Howlett and Mr. S. Mayall.

Also the following for permission to reproduce photographs: Rothamsted Experimental Station, U.S. Soil Conservation Service, Dr. I. Levisohn, Mr. B. S. Furneaux, M.Sc., F.G.S., and Mr. N. P. Burman, B.Sc., F.I.M.L.T., A.I.S.P. Thanks are also due to Mr.

Charles Birt, A.A.I., A.I.Arb., for assistance with some of my own photographs and some clerical work. Particular thanks are due to Mr. L. D. Hills for many helpful suggestions concerning the manuscript.

The official reports of our experimental stations, particularly Rothamsted, have been quoted extensively for the sake of up-to-date information. In some cases I have referred to experiments still in progress, and in such cases it should be borne in mind that too rigid conclusions cannot be drawn until their completion.

I

THE FOOD OF THE PLANT

'Many people have the concept that the nutrients in the soil are soluble, and that the plant puts its roots into the soil and sucks out nutrients along with the water. We've been in error to believe that might be the way it works.'
PROF. W. A. ALBRECHT, Prof. of Soils,
Missouri College of Agriculture.

The first creatures which might be truly called living on this earth were undifferentiated lumps of protoplasm, or living material, which derived energy for their life processes from such simple materials as were already available.

In the world of rocks, water and air there was no soil or humus or even plant nutrients as we know them, the only materials available were those exposed on newly fractured rocks or in watery solutions. Such organisms used oxides of iron, sulphates and ammonia, for instance, and they exist to the present time. Each was independent of its neighbour and provided all its simple needs, including its nitrogen. Eventually the chloroplast was developed: how, we do not yet know, but it was the beginning of life as we know it to-day. Energy derived from certain light wavelengths was utilized by these structures to combine carbon dioxide and water to form sugar (photosynthesis). This in turn was readily broken down to provide the organism with energy as required. It was here, no doubt, that the first symbiotic and parasitic organisms developed, greedy for sugar but not always willing to repay.

A further advance was the conversion of sugar to more insoluble substances—celluloses and lignins which provided the organism with a covering and enabled the first multi-cellular plants—the algae to evolve. For the first time there were energy-rich waste products litter-

13

ing the earth which, in turn, permitted another class of organism, the saprophytic (or 'living on dead matter'), to develop, a class dependent on another for its existence.

The next stage was the evolution of mosses which not only had the watery solutions from disintegrated rocks to feed on but also the dissolved and decaying remains of other organisms surrounding their 'roots'. As this vegetable litter increased, ferns developed with roots specially adapted to the absorption of materials from this early soil, subsequently followed by the flowering plants that we know to-day.

If it is true that evolution adapts an organism to its environment, it seems inevitable that these plants were influenced in their development by the presence of these decaying plant remains. From an evolutionary point of view it seems unlikely that a plant would not make use of the partly synthesized materials with which it is surrounded, thus giving it a lead over its neighbour, which has to consume rather more energy in building up its requirements from simple materials.

This evolutionary process occurred over many millions of years and it may be supposed that a similarly long period would be required to develop a soil to-day. One hears estimates of a thousand years to form an inch of fresh soil. This may be true where a cover of soil already exists but on bare, excavated or eroded ground, soil formation can take place in the course of a few years providing climatic conditions are favourable.

The reason for this speed of reclamation is not far to seek. The species of microbes, algae, mosses, and plants performing this work are already developed, ready to take their place the moment conditions become suitable. We do not have to await the slow process of evolution for organisms that can pave the way for plants—they are at work the moment the ground is exposed to the atmosphere and their activities proceed with increasing effectiveness.

Probably the finest example of the speed with which soils are formed and insoluble materials are rendered available, was the re-clothing of the island of Krakatoa, in the East Indies, with jungle, in ten years. Two-thirds of this island was blown off in a volcanic explosion. No trace of soil, or plant or animal life was left, over an area of several square miles. The island was quickly colonized by algae, then mosses and ferns, and in three years scrub growth was commencing.

Under temperate conditions the process is still comparatively rapid. One has only to note the speed with which old quarries and railway cuttings become clothed with vegetation though the exposed surface is sometimes bare rock. Wherever there is roothold, grasses, bushes, even trees, will grow. It may be that in some cases they are nourished by the overburden returned, or washed down by rain. But sometimes this overburden is 95 per cent subsoil—take, for instance, the great clay-pits of the London Brick Co. and Forders Ltd. at Peterborough, dug a hundred feet deep and abandoned twenty years ago and the Scunthorpe pit dug by the Frodingham Iron & Steel Federation—they are now full of trees.

Even granite in temperate climates, exposed to the atmosphere, has been shown to lose material at the rate of 17 lb. of silicon and 110 lb. of calcium per acre annually.[1]

It is often stated that there is no obvious relationship between soil fertility and biological activity. As the fertility of farming soils in many instances depends on the addition of readily soluble nutrients, such a statement is reasonable enough. The purpose of this book, however, is to demonstrate the practicability of self-support in agriculture by the utilization of natural processes, so the view is not applicable here.

It may well be that plants having evolved in the presence of simple inorganic solutions alongside the remains of living organisms they have the ability to use either or both with equal facility. On the other hand, it seems usual amongst the higher organisms generally that they are unable to supply all their needs from simple substances. There is little evidence as yet as to the needs of many higher plants in this respect, though it is certainly true of orchids and some shrubs and trees, and also fungi and much other microflora. But until plants have been grown in completely aseptic conditions it would be unwise to deny the possibility.

The most widely known association between plant roots and micro-organisms is that of leguminous plants and the various species and strains of Rhizobium, formerly known as Bacillus radicicola. These create galls on the roots of legumes in which they increase, extracting carbohydrate from the plant and in turn converting atmospheric nitrogen into the substance of their bodies. Some of these bacteria are digested by the plant and assimilated. At the end of the season the supply of carbohydrate falls off and the bacteria die, the nodule becoming detached from the plant, providing nitrogen for the following crop. The degree of infection depends on the nitrogen-carbohydrate ratio of the plant, so that by applying nitrogenous manures

[1] A. Demolon and E. Bastisse, *Soil Science* 1938, 46, 1.

to the soil, whether organic or inorganic, the nitrogen content of the plant is raised and the carbohydrate lowered in proportion, with a result that the fixation of nitrogen by the bacteria falls. As a general rule, then, the lower the soil nitrogen, the more is fixed by the bacteria. It is estimated that these symbiotic organisms require about 20 lb. of carbohydrate to fix 1 lb. of nitrogen.

The quantities fixed can be very large. Figures of up to 450 lb. per annum for wild white clover equivalent to 20 cwt. of sulphate of ammonia are quoted,[1] and my own observations on sweet clover left uncut for two years suggest the equivalent of 40 cwt. sulphate of ammonia, mostly in a gradually available form. Under present methods of agriculture, however, the amount fixed is between 50 and 100 lb. per acre for most crops.

Most leguminous plants fix more nitrogen than they require, particularly during long days when the night temperature is low, e.g. in May and June. It seems that the plant, which normally makes most growth at night, is hindered from doing so, while the organisms in the soil being at a more constant temperature continue their activity. This surplus nitrogen is excreted as an amino-acid, aspartic acid,[2] which appears to be the nitrogenous compound manufactured by the bacteria.

Many leguminous plants seem to have an excess of nitrogen all the time, excreting up to 50 per cent of that fixed. It is for this reason that there is generally a considerably greater total weight of crop when another kind of plant grows alongside the legume.

Take, for instance, the following table:[3]

	Yield of Dry Matter lb. per acre	Protein per cent	Nitrogen in Herbage lb. per acre	As Sulphate of Ammonia lb per acre
Kentucky Blue Grass	881	18	25	125
Kent W. W. Clover	3,072	35	156	780
Kentucky Grass & Clover	4,985	31	247	1,235

[1] E. B. Levy, Director of Grasslands Division, N.Z. Dept. of Agriculture; J. Farmers' Club 1949, 6.

[2] A. I. Virtanen and T. Laine, *Biochem. J.*, 1939, 33, 412–17.

[3] D. B. Johnstone, Cornell University. Referred to in *Jour. Soil Ass.*, Vol. 4, No. 3. For other references see W. R. C. Paul and A. W. R. Joachim, *Trop. Agricst.*, 1941, 98, 257; J. S. Papadakis, *Ann. Gembloux*, 1939, 45, 132; 1940, 46, 1; H. L. Borst and J. B. Park, *Ohio Agric. Expt. Sta. Bull.*, 513, 1932.

The Food of the Plant

The grass had taken up the surplus nitrogen from the soil so that there was no great amount left in the soil at the end of the season. If the clover had been grown alone, the total amount of nitrogen fixed may well have been considerably greater than that shown in the herbage, on account of the greater quantity of clover plants. This nitrogen would have remained in the soil and become liable to wastage, though much of it would be recovered in the succeeding crop. Also this one, as a comparison with various amounts of nitrogenous fertilizer:[1]

WHITE CLOVER AND YIELD OF GRASSLAND

Dry matter in lbs. per acre

Treatment	1936	1937	Total
Kent Clover	2,005	2,857	4,862
Ladino Clover	3,481	3,493	6,974
No Clover or Nitrogen	1,584	2,626	4,210
N_1	1,911	2,871	4,782
N_2	2,295	2,924	5,219
N_3	2,707	3,170	5,877

Ladino clover is a much stronger growing variant of white clover. Nitrogen was added in doses of 28 lb. ($=1\frac{1}{4}$ cwt. of sulphate of ammonia), N_1 one dose in April; N_2 one in April and one in June; N_3 one in April, one in June and one in August. The fertilizer used was 'Calnitro', which also contains calcium.

It will be seen that the presence of Kent white clover gave a rather greater yield than 28 lb. of additional nitrogen and the Ladino clover more than three separate doses of nitrogen, in spite of the ground it occupied at the expense of the grass. In addition, it will have provided the soil with some organic matter which the fertilizer would not.

Grain-bearing legumes, such as peas and beans, do not usually leave very much nitrogen in the soil. Most of it is harvested in the seed, as shown below.

[1] B. A. Brown, *J. Amer. Soc. Agron.*, 1939, 31, 322.

The Food of the Plant

AMOUNT OF NITROGEN FIXED BY LEGUMINOUS CROPS AND
THEIR INFLUENCE ON A FOLLOWING CROP[1]

Mean of two experiments, five courses of legume-cereal (rye or
barley) rotation

Crop	Nitrogen Harvested in Legume lb. per acre	Nitrogen Harvested in Cereal lb. per acre	Gain or Loss of Nitrogen in Soil per Rotation lb. per acre	Yield of Cereal cwt.	Total N. Fixed lb. per acre
Lucerne (as annual)	299	66	122	23·2	487
Clover	125	51	115	19·4	291
Sweet Clover (as annual)	170	51	84	18·9	305
Soya Beans	176	29	—8	11·8	197
Field Beans	103	25	—20	10·6	108
Cereal every year	—	22	—10	8·7	—

The amount of nitrogen fixed by various legumes seems to depend on
the circumstances under which they are grown. Experiments in Tunis[2]
showed that peas, beans, vetches and lentils fixed between 400 and
500 lb. of nitrogen per acre annually and left about 200 lb. in the
soil. While soya beans in prairie soil only fixed 20–25 lb. annually—[3]
probably because prairie soils are usually already rich in nitrogen.

A 'productivity index' of the effect of certain crops on the succeeding crop shows that, using the crop following a grass ley as the
standard, a crop of lucerne or sweet clover will increase the following
crop two and a half times as much over ordinary arable conditions.

Nitrogen is also fixed by free living organisms. Blue-green algae
are an important source in the Tropics, as it seems they only thrive in
warm, moist conditions. There are, however, other organisms at work
in temperate climates. Several types are recorded,[4] though Clostridia
and Azotobacter have been the most widely studied. A pasture in
Pennsylvania free from leguminous plants and Azotobacter fixed 15–
30 lb. of nitrogen annually. While two pastures at Rothamsted
allowed to run wild gained 92 and 60 lb. annually for twenty years,

[1] T. L. Lyon, Cornell Agric. Expt. Sta., 1936, 645.
[2] L. Yankovitch, Ann. Serv. Bot. Agron. Tunisie, 1940, 16, 303.
[3] A. G. Norman, Proc. Soil Sci. Soc. Amer., 1947, 11, 9.
[4] See S. A. Waksman, *Soil Microbiology*, 1952, p. 193.

the former, however, contained 25 per cent of leguminous plants. No record of the amount lost in the drainage is given, but it is certainly small.

While comparatively little is known about the variety, food requirements and numbers of organisms that fix nitrogen, a few factors are important. Abundant air supply for most of the time at least, an energy supply (organic matter) and certain minerals. While most of these organisms need also calcium and phosphates and traces of iron, copper and zinc, which are rendered more easily available to the organisms by the presence of humus, their activity depends on soil temperature so that fixation gradually increases until the end of August, and then rapidly declines. If there is no crop growing in the soil or there is no readily utilizable form of carbohydrate in the soil so that other organisms may convert it into protoplasm, this nitrogen, or some of it, accumulates until the rains of autumn wash it away.[1]

Azotobacter can fix nitrogen while living symbiotically with some bacteria under partially anaerobic conditions. Under such conditions the celluloses, which it cannot normally use, are converted to lactic, butyric and acetic acids which it can use.[2] This sort of thing occurs under rather wet conditions[3] and in compost heaps. The nitrogen incidentally is excreted as amino-acids or proteins which are quickly taken up by other organisms. Though, should anaerobic conditions be too intense or prolonged, there is a risk of denitrification by other organisms.

It has been claimed that the presence of plant roots assists the activities of Azotobacter perhaps by providing them with food or growth substances. A case has been recorded where the fixation of nitrogen was increased eight times by the presence of radish, rice, wheat and sorghum roots under laboratory conditions.[4]

As with the symbiotic bacteria fixation is depressed by the presence of soluble nitrogen, so that a crop should be kept growing on the soil to keep the nitrogen absorbed.

An important source of nitrogen, and to some extent phosphate and other minerals for the plant, particularly under natural conditions, is the soil humus and decaying organic matter.

Humus is a complex mixture of carbon-nitrogen compounds with

[1] For diagrammatic representation, see E. J. Russell and A. Appleyard, *J. Agric. Sc.*, 1917, 8, 385.

[2] E. H. Richards, *J. Agric. Sci.*, 1939, 29, 302.

[3] H. L. Jensen, R. J. Swaby, Proc. Linn. Soc., N.S.W., 1941, 66, 89.

[4] B. N. Uppal, Proc. Indian Acad. Sci., 1947, 25b, 173.

a ratio of carbon to nitrogen of about 10 to 1. It consists of bacterial protoplasm and its residues and also complex compounds derived from lignin—the most resistant of plant remains to decomposition. It also contains 1–2 per cent of phosphates and small quantities of potash and other minerals required by plants. A third to a half of soil nitrogen in humus consists of proteins[1]—probably the bodies and remains of nitro-organisms.

In virgin soils of temperate climates—particularly the dry ones—it forms a very large reserve of nitrogen. There is often over a 100 tons of organic matter dry weight to an acre, about half of which is carbon and one-twentieth nitrogen, i.e. 50 tons and 5 tons respectively. The wheat lands of North America and Russia have given crops for very many years almost entirely from this source. In some places it was eventually exhausted with unfortunate results for the farmers.

In hot and damp climates, on the other hand, this humus does not accumulate to any great extent except under water-logged conditions as it is broken down by micro-organisms much more quickly. In such areas plant food, especially nitrogen, is released as plant and animal remains are actually decaying, and there is no great reserve in the soil. Indeed, there is no need for it as plant growth and decay is continuous in such climates, such organic matter as exists is mainly as a small quantity of surface litter, perhaps 2 tons to the acre, with a few tons in the soil itself.

Some at any rate of this nitrogen is supplied to the plant in a readily available form as nitrates by the action of nitrifying bacteria —species of Nitrosomonas. One of the by-products of the breakdown of soil humus is ammonia, derived from the protein of soil micro-organisms. Normally, i.e. under natural conditions where plenty of energy food is available in the form of dead plant remains, this would be re-incorporated into the micro-organism unless and until no further plant remains were available. The production of ammonia, therefore, is likely to be greatest in spring and summer, as plants are growing rather than dying at that time.

The ammonia is converted into nitrates by nitrifying bacteria at a time when the nitrates can be best used by the plant. During the course of nitrification further organic matter is broken down by the nitrifying organisms to provide themselves with energy. Continuous nitrification therefore depletes the soil of humus and is one reason why the use of ammonia compounds as fertilizers can be detrimental.

[1] J. M. Bremner, *J. Sci. Fd. Agric.*, 1952, 3, 497.

Indeed, if some old experiments of Boussingalt, done in 1873 on humus-rich soil are any guide, the nitrification of a pound of soil nitrogen from humus involves the destruction of some 30 lb. of soil humus. It certainly cannot be less than 20 lb., as the ratio of carbon to nitrogen in soil humus is 10:1. This will be discussed more fully in Chapter IV.

It has been contended that a plant receives all its nitrogen in this way. Other reasons apart, this seems unlikely, as the nitrifying organisms need carbonaceous matter. This inevitably post-dates their appearance on the evolutionary scale to that of the lower plants at least.

There is another means by which plants feed, and that is by a symbiotic relationship with fungi. It is apparent that some 80 per cent of flowering plants have this relationship either temporarily or permanently and generally the healthiest and most robust plants have the greatest fungal development. These fungi seem rather sensitive to soil conditions, being poorly developed in the presence of abundant plant nutrients and also under water-logged and other toxic conditions.[1] It seems best developed under natural conditions when there is a plentiful supply of organic matter but with little soluble nitrogen in the soil. There are two types; in one, the ectotrophic, the fungal mycelium invades the plant roots, deriving nourishment partly from the soil and partly from the plant. This mycelium is then digested and absorbed by the plant. Here is an obvious means by which a plant absorbs protein and accessory food factors direct.

Under natural, competitive conditions the fungus by its great surface area has a much greater power for absorbing water and nutrients and would benefit the plant in this respect also. This may account for its universal development in trees which transpire enormous amounts. A large oak tree may use 200 gallons daily. It would also be of great benefit in those climates where transpiration is high and soils shallow, as instanced by Sir Albert Howard in the tea and coffee plantations of the Tropics. It is also able to compete for food more readily with the soil micro-organisms and can utilize some of the soil organic matter more easily than the unaided plant root.

The value of these fungi to trees under some conditions is well exemplified by Prof. M. C. Rayner's work on the trees of the acid peaty ground at Wareham Heath, Dorset (*Problems of Tree Nutrition*). Forest trees refused to grow there until composts were applied to the soil. The fungus-infected roots or 'mycorrhiza' refused to

---

[1] W. Neilson Jones, *J. Agric. Science*, 1941, 31, 379.

develop until supplied with suitable food and then developed rapidly, apparently destroying the conditions which inhibited their development in the first place and have continued to do so without further treatment.

A similar instance concerning crop plants, though whether mycorrhiza were in fact involved was not investigated, concerns a boggy piece of land drained by the Hampshire W.A.E.C. and farmed by Mr. A. R. Wills. The field was divided into three strips; one dressed with lime, on which crops and weeds failed to grow, and another with farmyard manure which gave patchy results. The last strip treated with an organic compost grew excellent crops.

The presence of organic matter helps some forest trees to grow on calcareous soils, through the aid of the mycorrhiza. They are apt to suffer from yellowing of the needles due to potash or magnesium deficiencies.[1] Rothamsted found this occurred after top-dressing young spruce seedlings with nitrogen, which would, of course, depress the activity of the fungus. They got over the difficulty by spraying the soil with sulphuric acid and adding inorganic fertilizers. I had the same problem with Sitka and Norway spruce transplanted at about 15 in. high on calcareous soil. For two years the Sitka spruce failed to grow more than 1 or 2 in. and were pale in spite of being kept perfectly clean and given nitrogen. They were left the second summer to get very weedy, chiefly grass. The following season they grew 18 in. and 2 to 3 ft. annually thereafter, without any further fertilizers. Grass seed was then sown amongst the plot of Norway spruce and, though they had remained stationary for three years, put on a foot of growth, and 18 in. to 2 ft. annually thereafter. A good crop of toadstools, presumably from the fungus, appeared twelve months after sowing the grass and has appeared annually since.

The other type of mycorrhiza known as endotrophic appears to live entirely within or on the surface of plant roots and does not appear to ramify in the soil. It is, however, more widespread in distribution than the ectotrophic; most crop plants and herbaceous plants seem to possess them and they are best devloped in those plants which are most robust.[2,3] They have so far been very little studied, but their needs appear to be much the same as the ectotrophic types.

It is interesting that a very fertile soil containing easily available

[1] Rothamsted Expt. Sta. Rep., 1951, 43.
[2] M. C. Rayner, *Problems in Tree Nutrition* (Faber and Faber), p. 131.
[3] A. A. Hildebrand, Canad. J. Res., 1934, 11, 18; 1936, 14c, 11.

nitrogen gives a larger plant but with a more poorly developed mycorrhiza. It seems that the fungi are only stimulated to develop on the root when it is rich in carbohydrate (or sugary materials) and poor in nitrogen[1]—the fungi utilize carbohydrate from the plant roots and give nitrogen and other nutrients to the plant, which they have extracted from the soil. (About half the nitrogen is in the form of protein with 10 per cent of amino sugars.) Thus, where the plant can obtain easily available nitrogenous nutrients it does not need the help of, or stimulate the fungus and uses all its carbohydrates in its own tissues. A fertile soil, on the other hand, has a high bacterial and fungal population, so that it is probable that any possible materials the plant misses from the mycorrhizal fungus it may obtain from the soil bacteria and fungi. In nature, however, a soil containing easily available nitrogen is something of a rarity, to say the least.

Though it may appear desirable, at first sight, for the plant to use all its carbohydrate itself rather than give 20 per cent or so to the mycorrhizal fungus in return for its supply of nitrogen and other nutrients, it may not in fact be so. When readily available nitrogen and other fertilizers are added to the soil a variety of factors, mostly harmful, come into play. These will be discussed later (Chapters IV and V).

When a soil is made rich in nitrogen by the addition of artificial manures, organic manures high in nitrogen, or sterilization,[2] the mycorrhizal fungus is not stimulated on account of the high proportion of soluble nitrogen absorbed by the root. The soil fungi and bacteria have only a small supply of organic matter ('energy foods') to live on and cannot develop freely. The plant will therefore not be obtaining the kind of food on which it has been evolved. It may not be properly nourished and this would account in some measure for a susceptibility to disease. It may not for instance be able to build up its proteins of exactly the same proportions or chemical structure as those provided by the fungus.

The need for mycorrhiza or other root associations may be supported by a recent Rothamsted experiment in which a variety of plants, including sugar beet, French beans, cabbages, barley, brussels sprouts and tomatoes, were sprayed with nutrient solutions containing nitrogen, phosphate and potash. The barley and others absorbed the nutrients and increased in size. The tomato plant grow-

---

[1] E. J. Bjorkman, *Medd. Skogsförsöksanst*, 1941, 32, 23; and *Svensk. Bot. Tidskr*, 1944, 38, 1.

[2] For an instance of sterilized soil, see Rothamsted Expt. Sta. Rep., 1948, 38.

ing in rather poor soil and showing obvious signs of starvation absorbed these nutrients, as proved by analysis, but did not show any increased growth. Now the tomato plant does not form mycorrhiza, at least not in this country.

Perhaps a mycorrhizal fungus was able to convert the inorganic nutrients into something that the sugar beet and barley could use—while the tomato was not so fortunate. Perhaps the tomato depends on other soil organisms to do this work, and these, of course, were short-circuited. It would be most interesting to see whether a tomato, in its native country and with the opportunity for mycorrhiza to develop, would respond to sprayed nutrients. For this reason, it is even more interesting to learn that in Florida—not far from its native home—the tomato does respond to sprays of nitrogenous fertilizer (urea) as much as to a similar quantity of urea or other nitrogenous equivalent added to the soil, but only when the soil is low in nitrogen.[1] This effect may, of course, be simply due to the kind of fertilizer used.

This possibly is a far-fetched hypothesis, but given a wider range of plants it could easily be tested. French beans also showed no increase in growth but as extra nitrogen in the plant depresses the fixation of nitrogen by leguminous plants this is not so surprising.

The case of cabbage and sprouts increasing in weight although not mycorrhiza formers is hard to account for on this hypothesis. They, however, are developed from native plants and may have never had the mycorrhizal function or have developed other means of converting inorganic nutrients. It has been amply demonstrated that one member of this family that has been studied, rape, has a quite extraordinary effect of increasing numbers of radiobacter[2] which live in association with plant roots.

Rothamsted have also started to experiment on the possible need of crops for mycorrhizal associations. Plants of wheat and clover were grown in pots, and it was found that the plants with mycorrhiza showed no increase over those without. It appears, however, that the plants were grown in sterilized sand and fed with a chemical solution of nutrients. In view of the fact that fungi require decomposable matter from which to obtain part of their food—a fact repeatedly stressed by Sir Albert Howard in connection with mycorrhiza, the experiment has not much value in this connection. It is not unex-

---

[1] J. Montelaro, C. B. Hall and F. S. Jameson, Proc. Amer. Soc. Hort. Sci., 1952, 59, 361–5.

[2] R. L. Starkey, *Soil Sci.*, 1938, 45, 207–49.

pected, therefore, that the mycorrhizal plants showed no benefit over the others. It would not have been surprising, in fact, as the fungus had to subsist on the plant, if the mycorrhizal plants had been smaller.

Some plants have never been recorded as having a fungal association in this country. The only crop plants in this category so far known are the potato, tomato, and some cruciferous plants including brassicas. It will be noted that these three are about the greediest crop plants we have. They must have plenty of easily available plant foods to make satisfactory growth. Perhaps it is because they are naturally (or artificially) adapted to a rich soil. But it is a fact that these plants seem to grow well with little or no organic matter in the soil, providing plenty of easily available nutrients are applied and air can get to the roots. Their resistances to diseases under these conditions are quite a different matter. The potato, of course, has a peculiar advantage in that it has a very large store of nourishment in the 'seed', which probably contains enough trace elements and other necessaries to see it through one generation at any rate. It can also develop quickly and take advantage of much of the artificial manures added before they are lost through various causes. Cases of good crops (in weight) annually, of early potatoes, from soils receiving nothing but artificials for many years are common. But fresh seed is used every year, or nearly every year, and the crop is harvested before blight is prevalent.

Of the dozen or so minerals required for plant nutrition the most important quantitatively are phosphates and potash. It has been assumed, with good reason, that these minerals pass into the soil solution and are absorbed into the roots by osmosis. There is no doubt that this does take place, but recent work shows that it is not the whole story. The root is selective: it has been shown, for instance, that a root can extract the phosphorus from potassium phosphate and leave the potash behind if it already has a sufficiency of that element.[1] Or even exchange one nutrient for another in the soil.[2] Absorption also depends on the root's content of reducing sugars (glucose and fructose) but not sucrose.[3] It is evidently not osmosis at work here but appears to be some kind of chemical exchange. Barley and pea roots were used in these instances—it may be that other plants have yet other mechanisms.

---

[1] Rothamsted Expt. Sta. Rep., 1952, 66.
[2] E. C. Humphries, *J. Expt. Bot.*, 1951, 2, 344–79; and *J. Expt. Bot.*, 1952, 3.
[3] Rothamsted Report, 1952, 66; E. C. Humphries, *J. Expt. Bot.*, 1951, 2, 344–79; and *J. Expt. Bot.*, 1952, 3.

What is more, roots can even excrete potassium, calcium and phosphates. Perhaps to the benefit of other plants in the vicinity. Certainly this will have an effect in altering soil acidity in the region of the root.

While a plant may be able to absorb much of its needs from the soil solution it is obvious that all the phosphates, for instance, cannot be absorbed in this way as they only occur as 1–3 parts per million of the soil solution. The amount of phosphate taken up by a crop of 18 tons of turnips per acre is about 25 lb. which, at a concentration of 3 parts of phosphate per million, would necessitate the absorption of some 4,000 tons of water, or 40 in. of rain. The crop usually gets only a quarter or third of this quantity during the growing season and, in fact, such a crop would only transpire between 400 or 600 tons. On this basis, the use of soluble phosphate fertilizers seems ineffective, except perhaps on clay soils, where phosphates seem to be absorbed by clay particles and readily given up.

Some soil minerals and clays will absorb phosphates which can be replaced by citric and oxalic acids[1] which are, incidentally, common constituents of plant sap.

A further means by which plants may absorb nutrients is by the principle of 'base exchange' or 'exchangeable ions', whereby an unwanted element in the plant may be exchanged for a needed one in the soil.

It has recently been shown that plants can obtain trace elements sufficient for their growth from such an extremely insoluble substance as ground glass by this means without their apparently passing into solution at all.[2] In fact, the principle has already been commercialized in the production of a glass powder containing all the known trace elements, for addition to the soil. This material even enables rhododendrons and other lovers of acid soil to grow in an alkaline soil. The iron which these plants cannot extract from a chalky soil is more readily available in this unlikely material.

The original source from which a plant obtains its minerals is by the breakdown and solution of soil particles—at least, under natural conditions. The soil solution and base exchange have been considered to be the only immediate sources of food supply for the plant.

It is, of course, undeniable that plants can be grown in solutions of food materials without any soil at all. They certainly can, but it is

[1] T. Kurtz, E. E. de Turk, R. H. Bray, *Soil Sci.*, 1946, 61, 111.
[2] Dr. F. L. Wynd, Michigan State Coll.

also true that these solutions are far from sterile and the roots are ensheathed by bacteria feeding on the sloughed-off cells and on the outwardly diffused cell sap.

If we could grow a plant in good health for several generations under completely aseptic conditions, it would seem at first sight to dispose of the need for microbes. A further point arises however. It has been suggested that the marked resistance of plants grown on naturally fertile soil to certain pests and diseases is due to the absorption of antibiotics and other substances produced by soil organisms.[1] It would, therefore, be necessary to find out whether this aseptically-grown plant was any more susceptible to disease than that grown in this kind of soil.

However, the real point at issue here is that one cannot compare the nutrient solution with soil. The mechanisms involved are quite different. A culture solution is not soil. It is artificially aerated, carefully balanced for acidity to ensure that all the constituents remain soluble, and carefully adjusted for strength and proportions of nutrients. No student of chemistry need be told what would happen if one tried to grow a plant in soil with the aid of a nutrient solution. In most soils certain of the ingredients would be precipitated and become insoluble. Provided aeration was attended to, the plant would almost certainly grow, but the odds are on the plant in the bottle, without soil, as far as yield is concerned if similar amounts of nutrient solution are used. However, more of this 'fixation' of nutrients by the soil later.

Plants can absorb fairly large molecules, larger than one would consider possible if all nutrients had to pass through the semi-permeable membrane of the root surface into the sap. That they can absorb fairly large molecules is evident from the use of the phosphorus insecticides which are absorbed into the sap through the leaves or roots. As is the phosphorus compound phytin and possibly the compounds extracted by the enzymes on the plant-root surfaces.[2] Large virus molecules added to the soil can infect the roots of tomatoes. Also carnivorous plants absorb the digestion products of their prey without apparently reducing them to simple inorganic compounds.

---

[1] For a review of the subject, see S. A. Waksman, *Bact. Rev.*, 1941, 5, 231; *Microbial Antagonisms and Antibiotic Substances*, 2nd ed., New York, 1947; and *Soil Microbiology*, 1952.

[2] H. T. Rogers, Proc. Soil Sci. Amer., 1940, 5, 285; *Soil Sci.*, 1942, 54, 499; B. R. Bertramson and R. E. Stephenson, *Soil Sci.* 1942, 53, 215; V. E. Spencer and F. M. Willhite, *Soil Sci.*, 1944, 58, 151.

There is certainly a movement in the other direction, too. Some plant roots can excrete amino-acids and vitamins of the B group.[1]

It may be that large protein molecules do not pass directly into the sap. There seems to be no reason why the products of microbial decomposition on the root surface should not be absorbed directly by the protoplasm lining the root hairs.

There is evidence that free decomposition products from soil organisms do not exist in the soil solution. An investigation by J. M. Bremner on the constitution of soil proteins,[2] found over twenty amino-acids in most sorts of soil by extraction with acid but none in the soil solution. It seems, then, that the death of most microbes must be violent, i.e. by immediate absorption by some other organism. As the eventual end of these materials is in the body of the plant it seems there is at least a strongly circumstantial case for the immediate absorption of these materials by plants also.

If, as has been supposed, the plant draws up its nitrogen, phosphorus and potash dissolved in the soil water, then the more rapid the transpiration, other things being equal, the more of these nutrients will be drawn into the plant—and the faster it will grow. In fact, this does not happen at all, the fastest growing plants are those kept in a damp humid atmosphere where transpiration is low. This is even true of plants grown in nutrient solutions. This apparent paradox may be resolved by the foregoing view.

It is interesting to note that many plants can absorb nitrogen compounds and minerals through their leaves much more efficiently than through the soil, the sugar beet for instance, is twice as efficient—practically all our crop plants show this effect.

In this connection a few pounds per acre of nitrogen and phosphate are deposited annually in this country from the atmosphere—while on the west coast the quantity of sea salt may be three or more tons per acre[3]—with the large variety of minerals it contains. Such quantities would seem harmful at first but they are, in fact, diluted with enormous quantities of rain; and chlorides, being readily soluble, pass quickly through the soil. This is distinct from wind-borne salt spray which can have devastating effects for miles inland.

Mr. C. Ingham, writing in the *South African Journal of Science*

[1] For examples: A. I. Virtanen, *Cattle Fodder and Human Nutrition*, Cambridge, 1938; P. M. West, *Nature*, 1939, 144, 1050; H. Lundegardh and G. Stenlid, Arch. Bot. 1944, 31a, No. 10.
[2] J. M. Bremner, *Biochem. J.*, 1950, 47, 538–40.
[3] N. H. J. Miller, *J. Scottish Met. Soc.*, 1913, 16, 141.

(January 1943), has endeavoured to build up a hypothesis to show that plants obtain the bulk of their minerals from the atmosphere. Be that as it may, there is one curious plant—the Spanish Moss of Florida—which certainly derives its nourishment from the air or air-borne dust. It grows on telegraph wires and tree branches. It seems likely that many of our lichens must do also.

There is no shortage of plant minerals in most farm soils, it is simply a matter of availability, and it is becoming more and more widely recognized that the key to this availability is organic matter. The problem seems to be partly chemical and partly biological. One element may have a strong affinity for another so that the plant is unable to break the compound. It is, for instance, known that in the presence of iron or aluminium—and most soils contain large quantities of these—phosphates are only extracted by plants with great difficulty or not at all. Recent work has shown[1] that the presence of five organic acids—citric, oxalic, tartaric, malic, and malonic—helped greatly in rendering or keeping phosphates available. All these acids are produced by soil organisms and are also common constituents of plant sap.

Probably the first official recognition of this aspect of availability in this country was by the research workers on orchard management who found that by grassing down the soil beneath the trees the various difficulties due to trace element shortages disappeared.

On a theoretical basis there is no need to add any artificial manures to our farming soils at all. By farming soils is meant those 33 out of 48 million acres in the United Kingdom which are considered as land capable of growing crops. This area contains about 50 million people. On a basis of a total of 15 cwt. of excreta annually per person, with a potash content of 0·21 per cent and a phosphate content of 0·26 per cent and allowing a further 25 per cent for minerals lost in discarded portions of vegetables and straw for industrial purposes, etc., we lose in our sewage and garbage some 100,000 tons of potash and 125,000 tons of phosphate annually. This is equivalent to the quantity taken from the land in food.

Taking an average of several farm soils,[2] there are almost 3·5 tons of potash and about 2 tons of phosphate per acre in the top 6 in. Assuming that animal dung and crop residues are returned to the soil eventually, there is still sufficient potash to feed all our population for over 1,000 years, and phosphates for 500 years, in this top 6 in. alone. Add a deep-rooting crop to the rotation, and unlimited reserves may be tapped.

[1] P. H. Struthers and D. H. Seiling, *Soil Sci.*, 1950, 69, 3, 205–15.
[2] (*See foot of p. 30.*)

In view of this it is difficult to understand why, in the year 1950–1, some 225,400 tons of potash and 423,300 tons of phosphate were added to the soils of Great Britain. About three and four times respectively the quantity needed to feed our population. In fact, our soils only feed half. The figures of 341,405 tons of potash and 559,465 tons of phosphate given as the 'maximum profitable level of consumption' by the F.A.O. seems even more absurd. Research on the manufacture and distribution of fertilizers seems to be out of step with that on its utilization.

Some of the minerals necessary to plants can be dissolved out by the weak acid solution of carbon dioxide in the soil—formed by the decay of organic matter and respiration of plant roots. The plant has other ways, too. Some plants are powerful extractors of minerals from relatively insoluble compounds. Lupins, for instance, excrete an organic acid from their roots[1] and can extract phosphates, so can lucerne and sweet clover. Quoting from my own experience a crop of sweet clover raised the phosphate and potash analysis of a plot of ground from low to medium, which represents quite an addition. Lucerne also extracts zinc and, in fact, can be used to correct this deficiency in the soil, while beets and allied plants collect quite a variety of minerals, especially potash and magnesium.

A more indirect way by which a plant can extract iron particularly is by the action of its fallen leaves on the soil. Thus a watery extract of ten different conifers, including Scots pine, aspen leaves, and to some extent ash leaves, has been found to dissolve iron and other minerals out of the soil,[2] often leaving it very acid for this reason.

Oats contain up to 0·09 per cent and broad beans 0·03 per cent of

---

(1) Analysis of various soils, total nutrients:

Soil	Nitrogen	Phosphate	Potash
Oxford Clay	·24–·50	·05–·13	·42–·90
London Clay	·34	·16	·43
Reading Loam	·41	·26	·45
	·38	·15	·37
Gravesend Loam	·13	·15	·30
Chalky Soil	·32	·30	·29
Boulder Clay	·20–·50	·07–·16	·49–1·05

(2) Analysis of rocks, per cent:

	K_2O	P_2O_5
Limestones	·33	·04
Sandstones	1·32	·08
Shales	3·25	·17

(F. W. Clarke, *Data of Geochemistry*)

[1] H. Schander, Bodenk., *Pfl. Ernähr.*, 1941, 20, 129.
[2] Rothamsted Expt. Sta. Rep., 1952, 56.

copper in their ash—an element found in very minute amounts in the soil. Buckwheat and broom are plants that grow in very acid soil and yet they contain large amounts of calcium.

Little work has been done regarding the possibility of making minerals available by extractor plants: it has generally been considered easier to apply them direct.

Many of the soil micro-organisms, too, have the power of extracting phosphates and presumably other minerals. Indeed, they must, or these elements would never have been brought into circulation in the early days of the earth's history. This point will be referred to again.

The oft-quoted results of an experiment by H. G. M. Jacobson and H. A. Lunt[1] showed that worm casts contained five times as much nitrogen, seven times as much available phosphate, eleven times as much available potash, three times as much available magnesium, and one and a third times as much available calcium as the soil of the top 6 in.

It seems however that this plant nutrient is derived from the plant litter eaten by the worm. As E. W. Russell[2] says: 'There is no evidence yet that the passage of inorganic soil particles through the worm's alimentary canal affects the availability of the plant nutrients they contain, nor is there any need to assume the existence of such an effect for, as Lunt and Jacobson point out, the worm casts are poorer in plant nutrients than the original litter.'

These are the figures with an average analysis of plant litter for comparison (available nutrients in parts per million of soil):

Plant litter p.p.m.		
(Garden wastes)	Cal.	2,500
	Mg.	1,500
	Pot.	5,000
	Phos.	1,500
	N.	5,000
	C.	80,000

	Cast	Soil 0-6 in.	Soil 8-16 in.	Approximate proportion of plant litter required to give the increased nutrient		
				Soil 0-6 in.	Soil 8-16 in.	Average
Cal.	2,790	1,990	481	1/3	9/10	2/3
Mg.	492	162	69	1/5	2/7	1/4
Pot.	358	32	27	1/16	1/15	1/15
Phos.	67	9	8	1/25	1/25	1/25
N.	3,530	2,460	810	1/5	1/3	1/4
C.	51,700	33,500	11,100	2/7	2/9	1/4

Figures for soft-stemmed and leafy vegetable material are quoted,

[1] *Soil Science*, 1944, 58, 367.
[2] J. W. Russell and E. W. Russell, *Soil Conditions and Plant Growth*, 1950, p. 183.

this being the kind of food normally consumed by worms—they pick out the softer parts of plant tissue, leaving the fibrous matter for further decay as cellulose and lignin are largely indigestible by worms.

It seems likely from the figures of additional nitrogen excreted, which could only have come from plant residues and microscopic soil organisms, that the worms ate about a quarter of plant residue with the earth. On this basis there should be a good deal more available potash and phosphate in the casts than there actually is—about four and six times as much respectively. So that although the worm cast contains much more available potash and phosphate than the original soil, in fact most of it has reverted to an insoluble form. The magnesium, however, seems to have remained entirely available, while there may have been an increase in available calcium. This perhaps is connected with the secretion of calcium carbonate in the worm's digestive processes, which coud be derived from unavailable calcium compounds. It may, of course, be merely that the plant litter eaten was higher in calcium than the figures I have quoted.

Even so it is likely that worm casts provide a useful source of food to plants and perhaps preserve potash and phosphate from reverting to an even more insoluble condition. These two nutrients normally rapidly revert to an unavailable condition. In addition the colloidal nature of the materials in worm casts renders them much less affected by the washing of rain.

Other figures given by A.P. Wiess[1] showed, in water extracts, 0·011 per cent of nitrogen in the soil, 0·014 per cent in the cast, and 0·00067 per cent phosphate in the soil and 0·015 per cent in the cast: a twenty-five-fold increase in phosphate.

There is another sense in which worms render soil minerals available to plants and that is by bringing subsoil to the surface in a fine condition mixed with organic matter. This gives micro-organisms an excellent opportunity to dissolve out insoluble minerals.

A very interesting experiment by H. Hopp and C. S. Slater[2] on the effect of earth-worms on clay subsoil showed the enormous benefit that these creatures can give to such poor soils. Barrels filled with clay subsoil, containing no organic matter, were divided into two sets of similar treatments, one having dead worms added and the other having living worms. There were three main treatments: in one the vegetation of the first year was returned to the surface. In the second it was all cut and removed and a covering of bitumen paper put over the surface as mulch—but without providing food for worms. In the third the vegetation was removed and no protection given. These

[1] *Journal of Soil Ass.*, Winter 1947–48. [2] *Soil Science*, 1948, 66, 6.

(a)

(b)

Plate 1

SOIL FORMATION BY NATURAL AGENCIES

*

Photographs taken of a quarry excavated a hundred years ago and untouched since.

(*a*) Trees and strong grass growing on the quarry bottom having formed 4–6 inches of stony soil from rock fragments. The trees are about 35 years old (Geological formation Carboniferous limestone).

(*b*) Showing grass colonizing bare rock, by accumulating fragments of rock and organic matter, preceded by mosses and lichens.

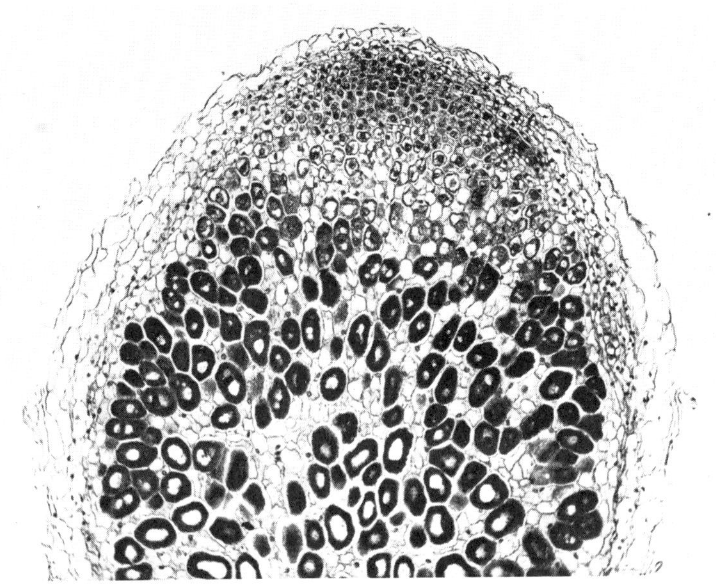

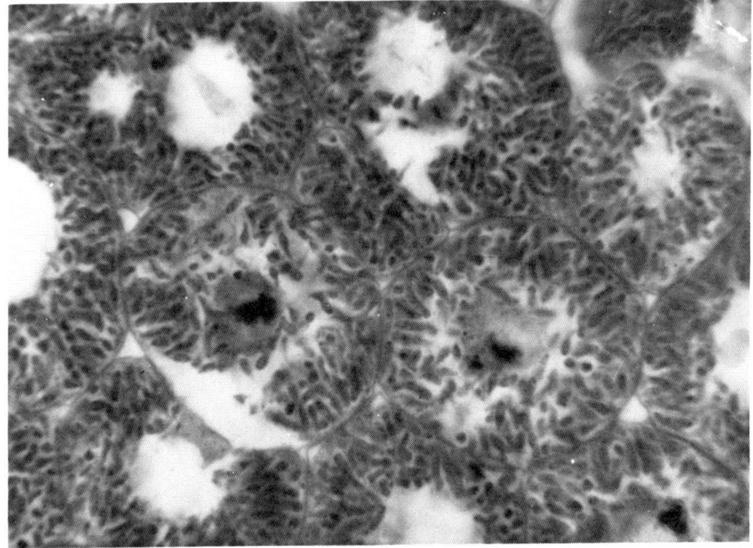

Plate 2

STRUCTURE OF A CLOVER ROOT NODULE

(*a*) Growing point of a clover nodule. The black areas are cells lined or filled with nitrogen-fixing bacteria which have been stained.

(*b*) Highly magnified portion of (*a*) showing masses of nitrogen-fixing bacteria in the cells. The large dark patches are the cell nuclei.

(By permission of Rothamsted Exp. Sta.)

treatments were further split into two—one-half being given a mixed fertilizer. There were then twelve different treatments, though some of these were duplicated in addition. These were started in autumn 1946 and the vegetation finally harvested in 1948.

The yields were enormously increased where live worms were added to the soil, up to four times, and even more so where fertilizers were added as well, though worms had much the greater effect separately. Returning the first year's growth greatly increased the second-year crop by providing the worms with much more food than the roots alone would provide. A complication arose through the appearance of clover which spread and had an appreciable effect on the final crop, presumably through fixation of nitrogen. An unexpected result was the appearance of a colony of ants in one of the dead worm barrels. These seemed to have the same effect as live worms on the plants, probably by aerating the soil and mixing in plant residues and their own excreta.

Over Winter Treatment	Fertilizer	Yield of Vegetation		
		Live Worms	Dead Worms	Gain
Vegetation returned	None	1·41, 1·47	0·29, 0·35	1·12
	Applied	1·64, 2·28	0·59, 0·61	1·36
Vegetation removed	None	0·55, 0·58	0·17, 0·34	0·30
Covered with paper	Applied	0·48, 1·34	0·43, 0·51	0·43
Vegetation removed	None	1·28	0·22	1·06
	Applied	0·64	1·22[1]	—0·58

A further experiment to test the infiltration rate of water, taken as the time for a given volume of water to pass through the soil surface, showed that the higher rates corresponded very closely with the higher vegetation.

Rothamsted have begun similar experiments and found that the more active worms were much more effective than sluggish kinds.

We have so far considered nitrogen, phosphorus and potash. Carbon hydrogen and oxygen we need not consider as the plant is surrounded by these in the form of water and carbon dioxide. There are also a number of minerals not necessarily employed in the actual structure of the plant but nevertheless indispensable as catalysts in many living processes and purposes we do not yet understand. These

---

[1] Affected by ants. Apparent loss in final column indicates that the ants were more effective than the worms under this treatment.

include sodium, magnesium, sulphur, chlorine, calcium, iron, copper, manganese, molybdenum, zinc, nickel, boron, silicon and fluorine. These are not all proved essential for all plants, though most at any rate appear to be. The difficulty is in identifying the small quantities that are sometimes required in some cases—1 part per million of molybdenum is sufficient to prevent 'whiptail' in cauliflowers—a deficiency disease. This plant can, in fact, take up sufficient of this element in the seed-bed, providing it is well supplied, to last it through its life. In some cases, too, sodium and chlorine for instance, certain members of the beet family grow much better with them—and yet it seems that they could do without them if necessary. A further complication is that some fifty or sixty elements have been found in plants, seemingly drawn up in the soil solution, and it is often impossible to say which of these are essential and which are not.

It is not the purpose of this book to give an account of the minerals required in plant nutrition, but a few examples will not be out of place.

Iron is required by plants and animals. It is dissolved out of soils under acid conditions, notably in pine forests, on peaty soil where it is often redeposited at lower levels where waterlogging occurs. On alkaline soils, on the other hand, some plants are unable to extract it —rhododendrons and heather notably. Organic matter helps to dissolve out iron for these plants.[1]

Copper, too, has a somewhat close relationship with organic matter. Peaty soil appears to hold it in an organic form and some plants have difficulty in extracting it until the peat begins to decompose by aeration. Good husbandry seems to be the answer. Some sandy soils, however, appear to be quite deficient and this appears to be one of those cases where the addition of small quantities of the mineral will rectify the trouble, without the risk of harmful effects.

Manganese is an example of an element that is required in a fairly definite concentration; too little retarding growth, while too much is poisonous. Sterilizing soil by heat, particularly if rich in organic matter, will often release toxic amounts, but under natural conditions organic matter has a 'buffer' action. Over-liming, on the other hand, will render it insoluble, as also depriving a calcareous soil of organic matter. Zinc behaves in a similar manner.

Boron, too, has to be present in a very precise amount—0·01 part per million causes deficiency symptoms, while over 1·0 part per million can cause toxicity. Here decaying organic matter is valuable again in

[1] J. Bonner, *Bot. Gaz.*, 1946, 108, 267.

its buffer effect, aiding extraction by the plant while at the same time reducing damage from an abundance.

Molybdenum is important, as it is necessary for the nitrogen-fixing bacteria of legumes. There is some evidence for a relationship between this element and copper, an excess of one creating a shortage of the other. It is, therefore, not always satisfactory to add one element to the soil as there is a risk of creating a relative deficiency in another. These relationships appear frequently in nature and they are one of the difficulties becoming obvious where much artificial manuring is used. Excess potash, for instance, will render phosphates and magnesium unavailable to the plant. These points will be discussed more fully later.

Silicon has the unusual property of assisting the plant to extract other minerals from the soil without appearing to be an important element of nutrition itself. It is present in widely differing amounts in different plants, but is especially found in the straw of cereals and grasses generally.

It should perhaps be pointed out here that humus itself is not a plant food, though it contains nutrients which are released by its breakdown. Humus in its scientific sense is the end product of decay for which the soil micro-organisms have little use—the indigestible remains of soil organic matter. Though they can use it slowly if nothing else serves. Plant food is mainly supplied by its release from plant remains during decay, and from the activity of bacteria on soil particles at the same time. Humus is synonymous with peat, at least the black solid fibreless peat defined as 'fairly stable amorphous brown to black material bearing no trace of the anatomical structure of the material from which it was derived.'

This is not to say it has no value: it has a good deal, partly as a moisture-holder and reserve of nutrients, but mainly as a 'soil conditioner', rendering it more easily worked and porous. It is, therefore, not humus itself that is of such value to the plant as food but the formation and subsequent breakdown of humus.

II

SOILS AND THEIR FLORA AND FAUNA

There are many different sorts of soils and many different conditions of soils. The first cannot be altered, the second can, and it is the farmers' object to maintain it as the most acceptable medium for plant growth.

In order that a plant's roots may thrive they need moisture, food, air, warmth and anchorage. An ideal soil, then, will hold and conduct moisture, contain an abundance of available food, permeated with an extensive network of cracks and channels, yet compact to aid water movement and anchorage. We have already studied the plant's food, but the remaining items are mainly bound up with soil structure.

Natural undisturbed soils consist of a friable surface layer filled with a mat of roots perhaps 2 or 3 in. thick and containing a quantity of decaying plant remains. Underneath there is another layer of crumbly earth of a dark colour about 6 in. thick, but in which the crumbs are all fitted compactly together like a jig-saw, being separated merely by fine cracks. Fine roots penetrate between these cracks and expand, squeezing the plastic soil crumbs into different shapes and, at the same time, closing up some cracks and opening others so that a constant supply of fresh feeding surfaces is available to the plant.

In addition to these cracks there are the larger perforations made chiefly by earth-worms, more especially on well-drained but moisture-retentive soils. These form the main conduits for the access of air and for the drainage of surplus moisture. Worms are constantly feeding on soil in one place and dumping it, in a more finely divided and mixed condition, in another.

In some parts of the world there are no earth-worms to provide drainage. Such places have long dry periods so that waterlogging is remote in any case. The need for these burrows, therefore, is not great, unless long wet periods intervene, as in the Tropics, where ants often serve the same purpose. In rather dry temperate climates, where microbial activity is low, a fibrous type of organic matter accumulates at the surface and its slow decay gives a finely granulated but fairly well aerated soil—the prairie soil.

Going deeper, the cracks become fewer but larger, so that the individual soil crumbs become soil blocks with more or less vertical divisions formed by the alternate drying and shrinking, and wetting and expansion, of the body of the soil. Down these cracks the plant tap roots go in their search for moisture and often for minerals also. The worm-holes, too, become fewer but larger, only the large worms are found below 9 in., the great majority of worms living mainly in the top 6 in. Disused worm-holes again become easy passages for roots.

These larger channels, in well-drained loamy soils, often go down 6 ft., and many of our crops go to this depth, especially root crops and others possessing tap roots. Roots of lucerne in some dry American soils have been found at 30 ft. There is seldom any need for roots to travel such distances for moisture in this country except on shallow soils overlying fissured rocks.

Plants use a lot of water while building up their substance, though only about $\frac{1}{2}$ to 1 per cent of the water taken up is actually built into the sugars and other materials that make up the plant body or used in plant processes. The remainder is evaporated by the plant through its leaves and stems during transpiration, drawing up a good deal of nutrients from the root at the same time.

It has been found that a plant needs at least 20 gallons of water to form 1 lb. of dry weight, and up to 80 gallons in dry climates—in the latter case simply to keep the plant tissues from wilting under the intense evaporation.

The water content of average soils on a rough basis is the equivalent of about 1 in. of rain for every foot of sandy soil and 2 in. for every foot of clay soil, with loam soils taking an intermediate position.

There are a variety of soils, more or less extreme cases, with high proportions of sand, clay or silt. A loamy soil containing about 40 per cent of sand, 40 per cent of silt and 20 per cent of clay may be taken as the standard, in which the sand and silt particles are bound together by clay to form the small soil crumbs. Only clay, of the mineral constituents, has this binding property—a sort of gelatinous stickiness which is only too obvious to those afflicted with clay soil. Silt, which is a grade of soil particles intermediate between clay and sand does not possess this stickiness. For that reason, it does not hold itself together in crumbs but easily breaks down to dust or mud with cultivation, and in wet weather effectively seals off soil pores with the possibility of suffocation of plant roots.

Sand particles, being coarser still, though having no binding properties again, will nevertheless allow air to penetrate into the soil. There is sometimes a danger, however, that if it contains a proportion of silt this will wash down and accumulate at a lower layer, forming a kind of pan.

All these extreme types have one thing in common. They are all improved, cumulatively, by decaying organic matter. They are improved in two ways. By an improvement in structure and an improvement in water-holding capacity. This is brought about by the presence of large numbers of bacteria and fungi and their remains, which are of a mucilaginous or gummy nature. This helps the silt and sand particles to combine into crumbs. Fungal hyphae have a similar effect.[1] It also permits the cracks in clay soils to remain open; this mucilage being of a much less sticky material than the clay itself. It creates lines of weakness between the clay crumbs and blocks. Unfortunately, these effects only hold as long as the micro-organisms are supplied with organic matter. A good soil with a good structure is one that will work to a tilth over a wide range of moisture conditions without the individual crumbs losing their entity under rain.

Water-holding capacity is improved in two ways: by the presence of these mucilaginous materials and, in the case of clay and silt soils, by the formation of fine cracks which hold thin films of moisture. Though the relative quantity of these colloidal materials is not great they can hold up to ten times their bulk of water. An annual dressing of 14 tons of dung per acre, over many years, has been shown to increase the water content of the top 9 in. of the soil by about a third, that is, from 2 in. to about $2\frac{3}{4}$ in.[2] And consequently reducing the loss by drainage, as the following figures from Rothamsted show.

Average number of days on which the field drains were running:

Dunged land   0·7 days ⎱
Undunged land   19·5 days ⎰ during the same period

The water-holding capacity of soils rich in organic matter is shown by the following instance. A 100 grammes of dried soil from each of the organic, artificial and mixed sections of the Haughley Experi-

[1] R. J. Swaby, *J. Gen. Microbiol.*, 1949, 3, 236–54.
[2] E. W. Russell and W. Balcerek, *J. Agric. Sci.*, 1944, 34, 123.

mental Research Farms was wetted with 50 c.c.s. of water.[1] The organically manured soil held the full quantity, the mixed section held 80 per cent, while the artificially manured held only 30 per cent of the water added. Soil containing 33 per cent moisture contains about 28 gallons of water per square yard in the top foot; this may represent a saturated moisture content from 10 to 60 per cent, according to the type and condition of the soil. As shown already, Rothamsted found that soil receiving annual applications of dung at 14 tons per acre held 35 per cent more water than unmanured land, though the volume of air spaces was little different. The extra water was therefore, it seems, held by the organic matter present and not in the spaces between the soil particles. It was also noted that the top soil where dung was used showed much less shrinking and cracking on drying than the subsoil and the unmanured soil. It should, however, be borne in mind that the use of farmyard manure is not a very effective means of raising the level of organic matter in the soil, as will be explained later.

According to recent work on the subject water moves very little in the soil, apart from drainage under gravity. Once surplus water drains away and the soil reaches its 'field capacity', then any loss of moisture in one part of the soil is not made good by movement from a moister area. This is undoubtedly true under the conditions of the experimental work, using a block of soil from an average arable field, low in organic matter. The same effect is amply demonstrated by ink on blotting paper—once the source of ink is removed the blot ceases to spread.

There is, however, a danger in applying the laboratory findings too literally.

It is a common observation, after cultivating a field and leaving it apparently dry in the sun, to find footprints and wheel-marks dark and moist the following morning. This cannot be condensed moisture from the atmosphere—the loose soil has a colder surface at night than the compact and would condense more moisture. The moist surface of the compact soil will, in fact, be losing moisture to the atmosphere as it is receiving heat transmitted from below—the dry, loose soil will not. There certainly seems to be some movement providing the soil underneath is moist, sufficient at any rate to germinate a seed—hence the value of rolling spring-ploughed soil. (In actual fact, some seeds need surprisingly little moisture for germination. Wheat, for instance, will germinate with a soil moisture content well

[1] *Soil Ass.*, 6, 3, 17.

below the wilting point of the adult plant.[1] Germination, however, is very much slower and the seed is much more prone to decay.)

There are a number of probable reasons for this. First, this matter of field capacity—the point at which the soil is holding as much water as it can without actually losing any by drainage. In practice, some soils seldom or never reach this condition as long as there is no crop growing in it—if they are high in organic matter. While Rothamsted soils drain off mainly in twenty-four hours with a slow trickle for a further few days, some American prairie soils continue to drain for months—even years. The following table shows this process going on for two and a half years. The soil was irrigated, and when water disappeared from the surface it was covered to prevent evaporation.

Sampling Dates	*3.10.34* *6.10.34*	*6.10.34* *17.10.34*	*17.10.34* *30.11.34*	*30.11.34* *8.5.35*	*8.5.35* *31.8.35*	*31.8.35* *22.1.37*
No. of days of drainage	3	11	44	159	115	510
From top						
4 ft. of soil	9·6 in.	1·5	1·6	0·2	2·7	0·4
9 ft. of soil	17·6 in.	4·1	4·0	0·8	5·3	1·3

By N. F. Edlefsen and G. B. Bodman, at Davis, Calif.[2]

It will be noted that water drains rather faster in summer than winter, due to a lowering of viscosity and surface tension with a rise in temperature.

While it is not expected that soils in the British Isles will approach this, except perhaps peat soils, it is evident that drainage of water is a very much slower process where organic matter is abundant. (Though, oddly enough, a drainage gauge 6 ft. deep at Rothamsted, filled with sandy soil kept bare of vegetation, gave a little water daily even in the dry season 1947 when the local drain gauges on clay soil had ceased for four months. Water is much more strongly held by clay colloids than by capillarity in sand.) This moisture is subject to capillary movement and will provide a plant's roots with moisture from a wider zone than would appear likely from laboratory observations.

It may be, too, that fungal hyphae in such soils will transfer moisture from place to place. There is no doubt that they can transport it for considerable distances—as, for instance, in the case of the dry-rot fungus.

[1] P. C. Owen, *J. Expt. Bot.*, 1952, 3.
[2] *J. Amer. Soc. Agron.*, 1941, 33, 713.

Soils and their Flora and Fauna

A practical example of this storage and movement of soil moisture is recorded by Dr. Martin Leake,[1] in which indigo plants are put in the ground after five months of Indian drought with a further three months of drought to follow. This is done by tilling the surface as soon as workable after the rains to prevent further evaporation, and, by keeping this tilth loose, subsoil moisture seeping up to the surface was unable to evaporate. Then, before planting, it was rolled repeatedly so that the conserved moisture was able to soak into the surface layers in sufficient quantity for the indigo plant.

Moisture is further conserved under natural conditions by the loose surface soil and plant litter, and of course the plant cover, preventing sun and wind from playing on the soil surface. It has been shown, at Rothamsted, that the evaporation from short grass is about 80 per cent of that from open water or wet surface in summer, and about 60 per cent in winter. They estimate that the evaporation from a tall crop would still be about the same as short grass. Evaporation, it seems, is much more dependent on heat received from the sun than other conditions. It has been shown that soil will lose $\frac{1}{2}$ in. of water by evaporation in five days in summer, but once the surface soil has dried out further evaporation becomes slight. Half an inch is quite a quantity when our summer months rarely give as much as 2 in., and that in small quantities at a time. If evaporation is slow, so that a dry surface skin is not formed, evaporation losses may account for as much as $1\frac{1}{4}$ in.—this would occur typically in early spring, and the evaporation of this quantity of water would also very considerably retard the warming up of the soil. These effects cease as soon as the soil is covered by a crop. A moist soil, provided it is well aerated, means that a plant does not have to seek far for its moisture, with the result that it makes a smaller root system than it would on a dry soil. This may mean quite a saving of energy and material for the plant.

It is not perhaps generally realized that it takes five times as much heat to evaporate a given quantity of water as it does to raise its temperature from freezing to boiling point. In fact, we only wish to raise the temperature of the water in the soil from its winter temperature of about 40° to about 60° F., that is about 20 degrees, not 180 degrees. We can say that the heat required to evaporate 1 gallon of water from the soil would enable 45 gallons of soil water to be raised to a temperature warm enough for any crop to grow in. Water comprises quite a high proportion of the soil weight and has a much

[1] *J. Soil Ass.*, Vol. 5, No. 2.

greater specific heat, from which may be deduced that from a quarter to a half of the soil's heat is held in the water.

Warmth of the soil is a matter of some importance as plant roots do not make any appreciable growth under 45° F. Rothamsted figures show that the soil temperature under turf is always slightly higher than bare soil except during the two hottest months, June and July. Although this difference is only about 1° F., its effect is cumulative—the faster a plant grows the more advantage it can take of improving conditions. The difference is due to the insulation effect afforded by the turf during winter and its effect in reducing evaporation. It has been shown already that evaporation from turf is less than from water or wet soil by 20 per cent in summer and 40 per cent in winter.[1]

The compact type of soil found naturally also warms through quicker in spring than a loose-cultivated soil, because the thermal conductivity of water and soil particles is about twenty-five and a 100 times respectively that of air. Thus a compact soil containing few air spaces—and those mainly in the form of vertical worm-holes which offer no resistance to the vertical movement of heat—will conduct heat from the surface to the lower layers where it is 'stored' and gradually released overnight. For the same reason the surface does not get very warm by day, a more equable temperature is maintained throughout the twenty-four hours. A spring-ploughed soil, on the other hand, contains many air spaces and conducts heat badly—the result is a hot surface by day, with a tendency to dry out, and a çold one by night—both conditions detrimental to seedlings. The subsoil warms only slowly.

Autumn-ploughed soil, likewise, will not be receiving warmth from below—this being the direction of heat flow in autumn as the soil cools down. Seeds sown on such land will have a cold surface on which to germinate, particularly in frosty weather, with a serious risk of lifting by frost.

Air in the soil is necessary for plant roots to breathe (and the micro-organisms). Generally speaking, they grow best with a carbon-dioxide content of less than 1 per cent, though they can tolerate 9–10 per cent for a short time. Should the soil surface be sealed by heavy rain—especially in warm weather—the carbon-dioxide content may rise tenfold during the day, and if prolonged will kill the roots. Under natural conditions, with a surface litter, sealing of the surface would

[1] R. K. Schofield and H. L. Penman, Institution of Civil Engineers, Sept. 1948, 75–84.

not normally occur and even under cultivated conditions, if organic matter is abundant, worms will quickly break up the crust.

The extreme forms of clay, silt and sand are liable to suffer from poor aeration for various reasons even under natural conditions. Clay subsoil is impervious and holds water in winter, especially if the land is flat. Worms cannot live in it—or at least the larger, deeper burrowing and drainage-providing kinds. Such land, therefore, tends to carry only surface-rooted plants which will not be so susceptible to waterlogging. As the clay dries out in summer the roots follow the moisture level but seldom penetrate more than a foot or two, for with the onset of wet weather these roots are suffocated and have to be renewed each year. One will not find plants with perennial tap roots or deep-rooted trees very successful under these conditions. Clay soils have usually immense reserves of minerals, but the conditions preclude their extraction by plants. If drained, however, they may become very fertile.

Silt soils are the hardest to manage under cultivation of any. Under natural conditions, providing there is no impervious layer below and rainfall is not more than the drainage can cope with, one finds a similar state of affairs as in a good loamy soil. Once the surface is exposed and surface litter reduced, the fine particles of which it is composed separate and work down into the cracks and wormholes, blocking them. Or the wind blows them away. Add to this the destruction of the cracks and burrows by cultivation, and one is left with a layer of material akin to fine flour, caked when dry and liquid when wet. Such soils must be given the minimum of cultivations and kept under grass as much as possible.

The sands may suffer from lack of air if they have much fine silt with them (and they very often do). Worms are not very abundant in sandy soils, so aeration is more dependent on the pores between particles and the rapid passage of water between them. This rapid passage of water may well cause trouble where decomposable organic matter is short by washing the finest particles of the soil into the subsoil, and under the acid conditions common in such soils a hard waterlogging pan of fine particles and iron oxide will form, very difficult indeed to break up.

A remarkable tribute to the drainage system of some undisturbed soils was paid by an article entitled 'Bottomless Forest?' in U.S. *Newsweek*, 6th November 1950. A large market garden had to dispose of a maximum of 10 million gallons of muddy water daily from its vegetable washing plant. An attempt was made to dispose of it by spraying it on a field of red clover, but after 2 in. of water had been applied the field became waterlogged. It was then sprayed on a patch of woodland adjoining at the rate of 50 in. a week. Pure filtered water ran out of the lower end of the wood. How long this drainage system will stand the strain remains to be seen, but there is no doubt that a natural soil structure is more stable than any artificially created structure can ever be.

Somewhat accidental support for the drainage system of un-ploughed soil came through a Rothamsted experiment.[1] They have been endeavouring to find out why the soil in a ploughed ley is so much drier at the surface than ploughed arable land. They found that it is due in part at any rate to the conductive capacity of the dead root fibres. The size of the conducting vessels was such as to provide strong capillary suction. This was true of ryegrass and several others but not of cocksfoot. While in dry weather they conducted water upwards.[2] They formed the interesting conclusion that to obtain the greatest benefit from these roots the turf should not be inverted but the roots cut off immediately below the plant. They propose to compare shallow rotary cultivation with ploughing.

Organic matter applied annually has a long-term effect which seems to be apart from the associated activities of micro-organisms, probably due to the humus products derived from lignin. It improves the friability or crumb structure permanently. At Rothamsted the proportion of crumbs larger than half a millimetre which stood up to wetting were increased from 28 to 55 per cent on one soil and from 54 to 70 per cent on a heavier soil by the annual application of 14 tons of manure per acre. The effect, however, seems to be less in dry climates.

Surface mulches have a very good effect on soil structure as all the benefit is concentrated at the surface. An experiment to test this benefit showed that the percentages of water stable crumbs over 1 millimetre of a soil were 3 on a bare surface, 17 where weeds and grasses were allowed to grow, and 34 under a straw mulch.[3]

Grasses, too, have a good effect, though it varies between species, apparently depending on the length and toughness and resistance to decay of the roots. The roots in effect create a mesh holding the soil particles together. There is in addition the fact that soil under grass is uncultivated and not subject to the losses of organic matter by cultivation, a factor that will be discussed in a moment.

[1] Rothamsted Expt. Sta. Rep., 1951, 31.
[2] Ibid., 1952, 36.
[3] R. E. Stephenson and C. E. Schuster, *Soil Sci.*, 1945, 59, 219; 1946, 61, 219.

Carrying the conception of evolution in relation to the plant a little further, let us examine the soil conditions under which plants have developed.

First, the bare rocks and water association of algae and mosses whose remains paved the way for ferns and higher plants. These, with the aid of their root systems and with fungi feeding on their decaying remains, bound together the finer particles of rock, dust and mud into a more or less granular structure, the soil. This in turn became the home of that ancient burrowing creature, the earth-worm, and many other smaller animals. During the development of our present-day plants, their roots lived in a more or less compact mixture of coarse and fine mineral particles and plant remains, interwoven with fungal threads, the whole being permeated with the burrows of earth-worms and soil fissures. The plant remains on the surface were dragged down bit by bit and consumed by the worms and the finely ground remains used to line their tunnels, while those worms living largely on soil brought their contributions of minerals to the surface in a more readily available condition for the plant.

The soil, being compact, conducted heat and water readily in all directions. To summarize, then, the plant had the following conditions:

1. Firm soil, giving good anchorage, and conducting water upwards from the subsoil to some extent.
2. Ready-made passages for the roots lined with concentrated plant foods, with ample air and moisture.
3. No risk of waterlogging in wet weather (providing the subsoil is not impervious).
4. Little risk of root damage due to heaving of the ground by frost.
5. Soil protected by a surface litter giving protection against rapid changes of temperature and moisture.

It is not generally realized how great is the extent and depth of plant roots. Isolated plants of wheat, rye, or wild oats were found to produce 40–50 miles of roots[1] under the dry conditions of Saskatoon, though when grown closely under field conditions only about $\frac{1}{2}$ mile per plant. This is still a lot of root to cram into $\frac{3}{4}$ in. of a drill. Such lengths would probably not occur under the moist conditions of the British Isles, though they would still be considerable, especially in a dry season.

The surface area of roots is also enormous. A cylinder of soil 3 in. in diameter and 6 in. deep in a crop of rye may contain 55,100 ft. of

[1] T. K. Pavlychenko and J. B. Harrington, *Canad. J. Res.*, 1934, 10, 77.

root hairs with a surface of 1,190 sq. in. Soya beans, on the other hand, may have a mere 1,960 ft. of root hairs and 43 sq. in. of surface. Plant roots of practically any crop will occupy the full extent of the top soil and often go down as far as soil conditions permit. Even lettuce will go to 18 in., strawberries 3 ft., pasture grasses to 6 ft. and sugar beet and other root crops to 6 ft. or more, where there is no 'panning'.

Soil that has its structure undisturbed by cultivation is not so likely to suffer from losses of nitrogen by leaching. Soluble forms of nitrogen are often held in the soil blocks, into which they are drawn in solution, as they become wetted by autumn rains which wash soluble substances down from the surface. As the soil becomes wetter the solution becomes more dilute until, by the time it begins to pass to regions below the root range, there is little soluble material in it at all. Most of the excess water from then on passes straight down cracks and worm-holes to the natural drainage system. Only the final drainings from heavy downpours contain appreciable quantities of nitrates.[1] Thus a good deal of nitrogen is stored over winter in the subsoil where it may be utilized the following summer—provided it becomes dry enough for the roots to reach these levels. This sort of thing cannot occur in a sandy soil, neither, as far as appearances go, where heavy winter rainfall is experienced. This storage does not seem to be very noticeable on my own soil, where forty or more inches of rain are the rule, nearly all in autumn and winter. Where it does occur it may account in part for the greatly increased crop after fallow.[2]

Another result of cultivation not generally appreciated is the loss of organic matter due to this factor alone. When soil is broken up a greatly increased number of surfaces are exposed to the air. Take a large block of soil and break it into several pieces. In addition to the surface of the original block there are several new surfaces. Each of these surfaces becomes the scene of increased microbial activity on account of the sudden access of air. The microbes live on the organic materials exposed until there is nothing usable left. This occurs with each ploughing and cultivation. This is all right if the organic matter can be replaced with large dressings of dung and compost, but it cannot be done these days. If the soil is undisturbed, bacterial activity and release of plant food will take place almost entirely on the surfaces of the cracks and earth-worm burrows which are at the same

[1] R. Warington, *J. Roy. Agric. Soc.*, 1882, 18, 1.
[2] E. J. and E. W. Russell, *Soil Conditions and Plant Growth*, 1950, p. 469.

time colonized by roots. It may be contended that an increase in surfaces would give the roots more scope for their activities. This is true, of course, but much of the organic matter is lost long before the plant can colonize these new surfaces made by ploughing or digging. It is better to let the plant make its own spheres of activity by splitting up the soil blocks with its roots as the need arises.

There is no doubt that digging and ploughing an undisturbed soil does produce larger crops for this very reason—but only for a time, a year or two perhaps, and then, unless large amounts of organic matter are added, there is a progressive decline.

Evidence of the loss of organic matter by cultivation has been given by most authorities in various parts of the world. The Fen districts, for instance, where intensely cultivated, lose from ¼ to 1 in. of soil per year. Such soils, of course, are very largely made of organic matter—peat, which, though a comparatively resistant form of organic matter, is being lost at an astonishing rate once drainage and cultivation permit the air to enter. This organic matter contains, and releases as it is oxidized, considerable quantities of nitrogen and other plant foods. Indeed, it has been shown that the use of nitrogenous manures on such soils is unnecessary and British fenland farmers could save themselves £25,000 annually.[1] Keeping the soil under grass considerably reduces the loss.

These losses are well shown in the following table after 30–40 years under cultivation (per cent).

Treatment	Cultivated Soil	Virgin Soil	Loss by Cultivation
15 yrs. in wheat, 15 in maize	2·74	3·89	29·5
Maize, oat, wheat	3·21	4·43	27·5
Lucerne last 12 yrs.	3·86	4·43	12·9
Lucerne last 7 yrs.	3·38	5·03	32·9
Lucerne last 7 yrs., badly blown 1 yr.	2·47	5·03	50·9
Orchard, clean cultivated last 7 yrs.	2·76	5·41	49·0
Grass last 5 yrs.	3·57	5·41	34·0
Heavily manured for 20 yrs.	4·68	5·05	7·3

F. J. Alway, Nebraska Agr. Expt. Sta. Bull, 111, 1909.

Particular note should be taken of clean cultivation where nearly half the organic matter was lost in seven years, and also of the effect of wind blowing in one season when 20 per cent of the organic

[1] F. Yates, *Nature*, 1952, 170, 138.

matter was lost due to this alone. Losses are shown to be smallest where the land has been under a hay crop for a number of years.

An example from the Ohio Experimental Station shows the losses under various rotations.[1]

Organic matter in top $6\frac{2}{3}$ in. (per acre)

	Carbon in 1,000 lb.	Nitrogen in 100 lb.
Initially (1894)	20·4	21·8
After 30 years of continuous maize	7·4	8·4
After 30 years of continuous wheat	12·8	13·1
Maize–oats–wheat–clover–timothy grass	15·5	15·5
Maize–wheat–clover	17·1	17·8

To convert carbon to organic matter, multiply by 2.

Maize, a clean cultivated, wide-spaced crop, has reduced the organic matter to about a third. Wheat to a much less extent, owing in part to the stubble and weed growth and close ground cover, giving protection from the sun's heat and reducing bacterial activity, but most of all to the fact that the surface of the soil is not disturbed by inter-row cultivations. The experiment also clearly shows the value of clover in keeping up the organic content. Another effect shown by cultivation is the loss of nitrogen which occurs—originally combined with this organic matter—shown in the previous table and again below.

LOSSES OF NITROGEN AFTER BREAKING UP PRAIRIE SOIL[2]

(From top nine inches)

	lb. per acre
Nitrogen present in unbroken prairie	6,940
Nitrogen present after 22 years' cultivation	4,750
Loss from soil	2,190
Recovered in crop	700
Deficit	1,490
Loss per annum	68

The annual loss is double the amount taken by the crops.

[1] R. M. Salter and T. C. Green, *J. Amer. Soc. Agron.*, 1933, 25, 622.
[2] F. T. Shutt, *J. Agric. Sci.*, 1910, 3, 335.

Plate 3

MYCORRHIZA IN A
CROP PLANT
(Sugar Cane)

(*a*) Active mycelium in the cortical zone.

(*b*) Digestion of the Mycorrhizal fungus by the sugar cane.

(*By permission of Dr. I. Levisohn*)

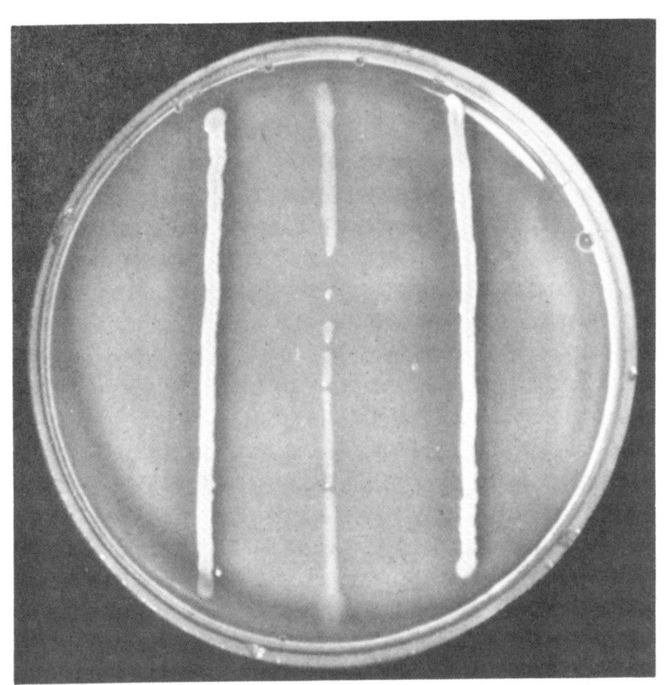

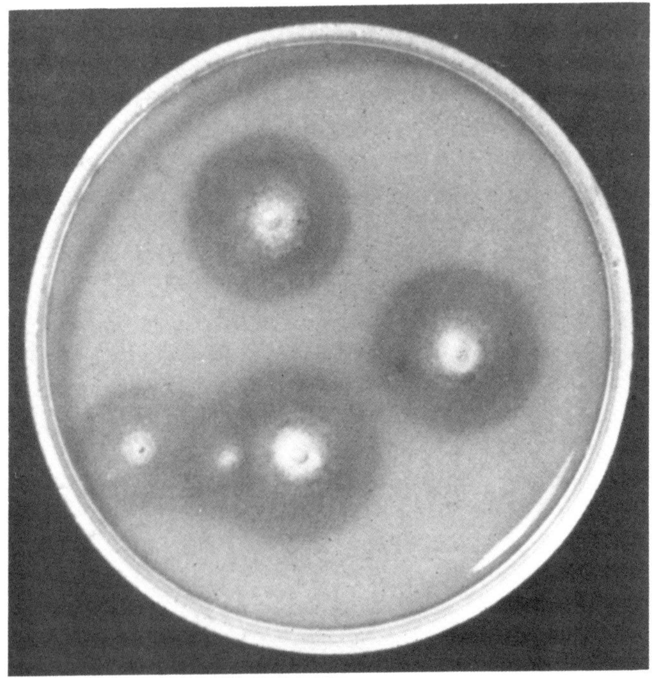

Plate 4

PHOSPHATE DISSOLVING ORGANISMS

(caption facing)

The basic reasons for cultivation of the soil have never, until recently, been questioned. Neither is there always a clear conception of its need and purpose. Ploughing originated when man first began to plant and sow crops: a Y-shaped stick or branch being used to break up turf and drag out roots, leaving a soil surface more easily managed. The application of power in the shape of draught animals was a natural development to increase the scope of his activities, until the refinement of the mould-board plough enabled him to bury completely all interfering material and growth. This provided, for the first time, a uniform, weed-free surface, permitting the development of mass-production methods and machinery. This, in turn, brought its own ecological association of stoloniferous and quick-growing annual weeds, with a demand for machinery capable of deeper burial, in the hope, though with little factual backing, that this would lead to cleaner land. There is also the view that deeper cultivation will provide a deeper root run and bigger crops, but the narrowness of this view has already been shown. There is, too, the not unimportant psychological element which provides the belief that the more effort is put into a job the better it will be. Much depends, as in all other work, on the application of the effort.

The need for a clean seed-bed remains paramount, but we have now implements at our disposal which will provide a clean, and finer, seed-bed without the necessity for deep cultivations—with a large saving in motive power required and consequent weight of machinery. Such implements comprise the rotary cultivators and to some extent the disc harrow. All the tilth a seed requires is sufficient to cover it, a fraction of an inch in most cases; and if firm, moist, but unpanned soil exists immediately below, it will not have to waste its energies reaching for daylight and moisture through several inches of soil.

Many will ask how one can incorporate manure and fertilizers using such shallow cultivations. It may be that the problem does not arise—this aspect will be discussed in Chapter IV.

Not only is the plant itself very dependent on soil conditions but

---

(*a*) Culture of micro-organisms in a medium containing only insoluble phosphates. The outer streaks contain phosphate-dissolving organisms, the centre streak does not and has been unable to make growth.

(*b*) A species of Penicillium growing in a similar medium, showing clear areas from which it has dissolved the phosphate. (The organisms in both cases were obtained from compost.)

(*By permission of Mr. N. P. Burman and the Soil Ass.*)

those organisms in the soil which help to maintain congenial conditions for the plant are also in need of moisture and air. Soil, particularly a fertile soil, is much more alive than a casual glance would show. We can see the larger insects and worms but the greater bulk of living things is invisible—consisting of bacteria, fungal mycelium and a variety of less well known organisms. Most of these are saprophytic—that is, they live on decaying materials. Others are parasitic, they live on their neighbours. Some derive nourishment from the soil fragments themselves.

The population varies according to the amount of organic matter present and the degree of moisture and warmth of the soil.

For bacteria, Rothamsted give numbers of about 3,000,000,000 per gramme of soil, weighing a ton to the acre, in the top 6 in. of a pasture. On arable land they amounted to about 3,500,000,000 where dung was used annually; 1,500,000,000 where complete artificials were used and 1,000,000,000 with no manure of any kind. These numbers vary surprisingly according to moisture and temperature, and even for no apparent reason at all, the numbers being doubled or halved in a matter of an hour or two. As these form the food of a large number of other soil organisms, an increase will be quickly followed by an increase in the predators and so a balance is maintained.

The food of bacteria is very varied, they live on anything from old iron to boot-leather. Some species are very simple in their tastes, living by breaking down inorganic substances—nitrates to give nitrogen, sulphates to give sulphuretted hydrogen or sulphur, and also on insoluble things like iron oxide and phosphatic minerals. It has been said that nothing on earth is immune to bacteria, and if it is a strain will soon develop. There are even some bacteria that live on carbolic acid in dilute solution, mothballs and salicylic acid, while another frequents petrol stations, living on tetra-ethyl-lead—the anti-knock ingredient of some brands. Some need more varied diets, and some, especially those living in rich soil, require quite a number of amino-acids, vitamins, and the like, for their growth, while some parasitic kinds will often grow only on one kind of food—their host. As a consequence, in a soil poor in organic matter, most of the bacteria are found surrounding the plant roots from which they derive their food, particularly those types requiring amino-acids[1] and vitamins, while, in the remainder of the soil, they are comparatively scarce.

Intermediate between these and fungi are the actinomycetes which

[1] P. M. West and A. G. Lochhead, *Soil Sci.*, 1940, 50, 409–20.

take on the appearance of either, according to circumstances. These appear to live entirely on organic materials but not a great deal is known about their activities at all. One is suspected of fixing nitrogen in nodules on alder roots. They thrive best under warm and comparatively dry and airy conditions, and are to be found commonly in the centre of rather dry compost heaps. These and fungi are the group chiefly responsible for the production of antibiotics in the soil.

The fungi live also on organic materials and, in addition to attacking cellulose and proteins, some also attack lignin, the woody material of trees and hard plant stems. It has been suspected that some may be able to use atmospheric nitrogen, but the evidence so far is not complete and most of them have to rely on organic or inorganic forms of nitrogen for protein manufacture and carbohydrates for their energy supply. It seems to be generally true of the fungi that, though they will grow in many cases on mainly inorganic materials, they grow faster and keep growing longer when organic materials are used. Thus *Penicillium notatum* of world-wide fame, which was first grown for penicillin in a solution of inorganic salts and sugar, was found to give a much bigger production of penicillin by the addition of a proportion of corn-steep liquor. Mushrooms, too, do not give so good a yield on composts containing a proportion of inorganic nitrogen as on plain horse manure or composts made with organic forms of nitrogen. It seems that traces of substances found only in the organic materials are necessary. There is, too, a connection between the species of fungi in the soil and the type of crop or vegetation, it has more influence on them than the rich food provided by farmyard manure.[1]

The fungal threads or hyphae worm their way through the soil by the simple process of growing at one end and retracting their contents from the other, leaving an empty tube, which in turn is eaten up by bacteria. Their numbers again are high and the number of pieces of hyphae vary from a few thousand to a million per gramme of soil, with a total length of up to 400 yds. in a garden soil.

A few words about the algae would be fitting here. These comprise the most primitive of green plants. They are extraordinarily diverse in structure from microscopic one-celled organisms to the largest seaweeds. The diatoms and filamentatious threads found in ponds and on damp earth also come into this class. They are of considerable evolutionary importance, being the first creatures to convert carbon dioxide and water to sugar by the aid of light, and it is from these

---

[1] J. Singh, *Ann. Appl. Biol.*, 1937, 24, 154.

first plants that the whole plant kingdom that we know to-day has developed. The enormous thicknesses of chalk and limestone rock are largely made up of their remains, and those plants and the shelled molluscs that fed upon them must have fixed prodigious quantities of carbon from the atmosphere, as chalk, in this way.

The soil types are mostly microscopic, consisting of one-celled, or groups of one-celled, types, and threadlike kinds living on or near the surface. They can extract much of their mineral needs from simple inorganic materials and those normally unavailable to plants, as shown by the instance of Krakatoa previously mentioned, where these organisms were the first colonists.

One group, known as the blue-green algae, are important fixers of nitrogen in warm climates; in fact, it seems that the rice crop, which supports a large part of the world's population, depends upon their activity for nitrogen and aeration of the soil. Whether any algae have this property in temperate climates has not yet been conclusively proved, but a symbiotic relationship between the green algae and a nitrogen-fixing organism, Azotobacter, has been suspected. However, the ways in which nitrogen is fixed in soils are very imperfectly known, but it is obvious that it is fixed in considerable amounts. The figure of 60 lb. per acre per annum for a non-leguminous pasture at Rothamsted is proof of this.

Their numbers vary enormously, they have been found in numbers from 100,000 to 3 million per gm. of soil. Generally speaking, they are more abundant in soils containing much nitrogenous matter. There is a popular fallacy amongst farmers and gardeners that a green surface to the soil normally indicates lack of lime. The result of an application is simply to reduce the algae and its nitrogenous matter to ammonia with a grave risk of loss. If, of course, the greenness was due to nitrogen from sulphate of ammonia, the soil will almost certainly show an acid reaction, which, in contrast to that shown by decaying organic matter (carbonic acid), is a true acidity and does need lime.

These surface algae are, in fact, the primary colonizers of bare eroded soils and are very important as such. They provide the first traces of organic matter without which most higher plants fail to get established. They have the power of forming a gelatinous sheath over the soil which has the curious property of permitting water to pass through it, thereby protecting the surface against erosion—an interesting example of evolutionary adaptation.

Besides these low forms of plant life, there is a wide variety of soil animals, starting with those forms akin to unicellular plants to the highly specialized insects.

The protozoa are mainly microscopic animals such as the Amoeba which ooze along the bottom of ponds and in damp soil, rather like blobs of treacle, others are free-swimming by means of cilia or flagella. They feed mainly on bacteria, but one or two—like Euglena —cannot be classified exactly as plant or animal as they sometimes possess chlorophyll like green plants. All of them exist only under moist conditions, turning to cysts or resting stages during drought. Their numbers, like those of bacteria, vary greatly over short periods, but are generally greatest in spring and autumn when warm moist conditions prevail.

Here is a table collected from various sources:

SOIL ANIMALS (IN MILLIONS PER ACRE)

Animal	*Manured Dung 14 tons p.a.*		*Unmanured*		*Old Pasture*	
	No.	*Dry Weight lb.*	*No.*	*Dry Weight lb.*	*No.*	*Dry Weight lb.*
Earth-worms	1·010	110·0	0·460	50·0	8·000	900·0
Centipedes & Millipedes	1·780	80·0	0·880	40·0	1·800	80·0
Springtails	2·390	0·2	0·590	0·05	54·000	4·5
Insects (inc. Springtails)	7·910	13·2	2·475	6·05	78·500	195·0
Spiders & Mites	0·970	2·0	0·347	0·05	4·100	9·5
Eeelworms	3·600	3·0	0·790	0·8	7·600	6·3
Live weight (x 4)	1,010 lb.		400 lb.		2 tons	

There is another, larger, group of soil animals, comprising nematode worms—eelworms, good and bad—earth-worms (about which more will be said later), mites, millipedes, wood-lice, and, of course, the insects and their larvae. These creatures play a very important part in the production of humus by chewing up plant and animal remains into fine particles, mixed with their digestive juices and soil. This mixture provides ideal food for fungi and bacteria which continue the breakdown, releasing plant food. The process is to some extent a descending spiral, a given particle of plant tissue may be invaded by fungi, secondary fungi, eelworms, wood-lice, earth-worms and back to fungi again, each extracting what it can and handing it on for further disintegration.

They are mainly active in the top 2 in. or so of the soil, needing well-aerated conditions. The earth-worms provide an exception, of course, spending most of their time in the top 6 in., but will burrow to several feet in dry or cold conditions.

Though these animals do much work in breaking down the organic matter, they actually only consume about one-fifth of it—most of them being unable to digest cellulose and lignin of which the greater part of plant litter consists.

Notice, in the table on p. 53, the enormous numbers of worms and springtails which feed largely on plant litter in the old pasture.

In order to obtain live weight, the figures should be multiplied by four, reckoning an average of 75 per cent water. So a very rough calculation gives half a ton of animals per acre in manured arable land, less than a quarter of a ton in unmanured arable, and 2 tons in old pasture. There is, in addition to this, about four times this weight in bacteria and fungi, etc., so that the quantity of living organisms in old pasture is something like 10 tons. These figures are collected from tables issued by Rothamsted and other national research institutions and may be taken as reliable. It may come as something of a surprise to the farmer to know that he has thirty times as much stock underground as on the field. There have been estimates of 40 tons per acre which, no doubt, are possible on very heavily manured soils and those rich in organic and decaying matter such as some market-garden soils.

The activities of earth-worms which, when present, outclass all other soil animals, need special mention. Charles Darwin was the first to give them any close study and wrote a book on them which is a standard work to the present day.

The large earth-worm, *Lumbricus terrestis*, often goes down 5 or 6 ft. in well-drained soils, while the two species which cast soil on the surface, *Allolobophora longa* and *A. nocturna*, go down about a foot. There are about a further twenty species in this country which burrow in all directions, mainly in the top 6 in.

They need a well-aerated soil and yet a damp one, so that they are not so common in heavy clays nor in shallow or gravelly soils. Neither will they tolerate very acid conditions and, therefore, are infrequent on peaty ground and heaths. This is probably connected with the fact that they use a good deal of calcium carbonate in their digestive tract. They are, in fact, sensitive to pH changes of only 0·2. They need, of course, food to provide energy for their activities and for body-building, but apparently can only digest the more readily available carbon compounds—sugar and starches—deriving their protein from the soil organisms eaten. Even so, they contribute considerably to the process of decomposition by grinding up the larger fragments into a form more easily utilized by other organisms. It has been shown, for instance, that they reduce the carbon-nitrogen ratio of a straw compost from 23 : 1 to 11 : 1 in two years, while the soil micro-organisms alone only reduced it to 18 : 1. It is likely, though, that some cellulose is digested by the bacterial action in the gut. Large populations, therefore, are only found in soils high in readily available organic matter and not too acid. Such soils are those covered by crops or their residues all the year round. For arable land, winter green manure or young weeds serve a useful purpose, and in the market garden, leaving vegetable tops and leaves on the surface assists in keeping up their numbers. Under such conditions populations as high as those under grass can be maintained.[1]

The larger species pull bits of dead grass and leaves into their burrows and excrete the remains at considerably lower levels, thus taking plant and bacterial foods down to the deeper regions of the soil.

The quantities of worms vary according to the suitability of the environment—arable soils low in organic matter contain perhaps half a million, while a well-manured one double this quantity. Old pastures on good loamy soil contain up to 8 million, and possibly more.

The weight may be anything from a few hundred pounds to 2 tons per acre. In fact, a well-stocked pasture may carry six times the weight of worms as farm stock. Their contribution to drainage is considerable. In my own soil, even at 9-in. depth, there are about 35 worm-holes to the square foot, i.e. $1\frac{1}{2}$ million to the acre. A field of irrigated lucerne has been shown to possess 6 million holes at the surface. It will be seen, therefore, that a good worm population is a great insurance against waterlogging. I know of adjacent fields on the same, rather heavy, soil, one covered with rushes, the other with good grass—and worm casts. The former carries stock in winter— the other has not carried any quantity of stock in winter for many years. The reason, I think, is simply that the puddled condition of the former soil in winter exterminates the worms, leading to a further loss of aeration and drainage. While many good farmers are aware of the

[1] H. Hopp and H. T. Hopkins, *J. of Soil and Water Cons.*, 1946, 1, 2, 85–8.

dangers of poached land, there are many more who are not, and even resort to expensive drainage schemes, when all that is required is better management.

The amount of earth moved annually is prodigious. Recent estimations of the amount of soil cast annually per acre in pastures by worms are between 1 to 25 tons on the surface and 4 to 36 tons underground. This results in a pore-space in the soil of between 40 and 67 per cent, according to the proportion of surface-casting species.[1]

Worm casts have a somewhat more stable structure than the surrounding soil, providing the worms have some organic matter to feed on.[2] When a pasture is ploughed up and converted to arable their numbers remain constant for about six months, when the turf will have rotted, and then their numbers fall quickly. When re-sown, their numbers slowly return, generally taking four to seven years to return to the original level.[3]

The amount of food consumed by these various soil organisms is naturally considerable. Rothamsted give figures to show how the organisms in a field, given an annual dressing of manure at 14 tons p.a., used the same number of calories daily as twelve men, the field producing food during the season for two men. An unmanured field used the calories of three-quarters of a man daily, giving food for half a man. Even this production of calories of heat daily would not very significantly alter the soil temperature. Imagine twelve men buried beneath the soil surface, spread out over a full-sized football pitch. They would get cold long before the soil got noticeably warmer.

While this may seem a great waste of food, it consists almost entirely of cellulose, lignin, and humus, which is inedible, as far as we are concerned.

[1] A. C. Evans, *Ann. Appl. Biol.*, 1948, 35, 1–13.
[2] R. J. Swaby, *J. Soil Sci.*, 1950.
[3] A. C. Evans and W. J. Mcl. Guild, *Ann. Appl. Biol.*, 1948.

III

ORGANIC AGRICULTURE

An outline of the conditions under which plants have evolved leads us on to the position of agriculture in relation to their evolution.

Agriculture goes back perhaps ten or twenty thousand years—quite a short period, evolutionarily speaking. Most agricultural developments have occurred only during the last two hundred years. It seems likely that the nearer our methods of growing crops approach the conditions under which plants were developed the more likely they are to thrive. Some of our plants, no doubt, have by artificial selection become more closely adapted to modern methods of agriculture to a very limited extent, but there is now a good deal of evidence from practical results to show that farming methods based on these evolutionary principles have a value distinctly above normal methods. Such evidence may be found in crop yields, long-term fertility, nutritional values of produce, plant and animal health and cash savings in cultivations. And, as a result, in the growing number of farmers adopting these methods.

Modern agriculture, briefly, is based on the plough and farmyard manure with, more recently, the addition of artificial fertilizers. The principle being the supply of readily available plant foods from outside sources and artificial maintenance of aeration.

The methods covered by the term organic, or biological, farming are, in principle, the utilization of natural processes to release plant foods, which are already present in the soil. This is achieved by supplying soil organisms with abundant energy-providing food in the form of plant and animal residues. Also by using cultivation methods which conserve these residues and minimize losses of plant nutrients from the soil.

A detailed account of these processes and methods will be given in Chapter IV, but first let us look at some of these practical results.

The immediate advantages to the farmer lie in the saving of the cost of fertilizers and a considerable reduction in the time and draught of cultivations. The making of compost, where necessary, must be set against this, but its distribution is considerably easier than the dung and plant residues of which it is composed. With careful management and good organization, by far the greater part of the dung will be dropped in the field.

Crop yields by organic methods compare well with those achieved by other farming methods.

The following table of yields at Haughley Research Farms, where a comparative test is being carried out, is a fair example of wheat yields.

YIELD IN CWT. PER ACRE

Year	Organic	Inorganic	Mixed
1941	16	$18\frac{1}{2}$	$16\frac{1}{2}$
1942	31	19	$26\frac{1}{2}$
1943	21	22	31
1944	$28\frac{3}{4}$	$23\frac{1}{2}$	26
1945	25	$17\frac{1}{2}$	15
1946	22	$18\frac{1}{2}$	$25\frac{1}{2}$
1947	15	9	14

Seed used is derived from its own section.[1]

Their figures for clover seed are outstanding, there seems to be some particular value for organic methods here. A crop was sown in 1944 and both fields yielded 30 cwt. per acre of hay at the first cut. The second cut was threshed for seed. The organic section gave 6 bushels per acre and the inorganic $3\frac{1}{2}$ bushels.

This seed was sown in 1946 and both fields yielded 30 cwt. of hay per acre the following year. The seed, however, turned out at 6 bushels and 4 bushels respectively. It is interesting in this connection that crops of field beans, another leguminous crop, do much better on farmyard manure than inorganic fertilizers. One can get bulk in a plant with inorganic nutrients applied to the soil, but there is a suggestion that something is lacking when it comes to seed formation, with its complex requirements.

Sir Edmund Neville, Bart., records how after disappointment with orthodox methods he turned to organic methods of farming—followed by a rise in production—and gives examples of a comparative test of yields on an experimental field. He gives his yield as follows:[2]

<hr>

[1] *Journal Soil Association*, Vol. 2, No. 2.
[2] *Journal Soil Association*, Vol. 4, No. 4.

Wheat

Compost only	23½ cwt. per acre
Artificials only	17¼ cwt. per acre
F.Y.M. and artificials	24¾ cwt. per acre

The use of 7½ cwt. of fertilizer with the F.Y.M. produced 1¼ cwt. more than compost alone—scarcely justified, without considering the condition of the ground for next year's crop (see Chapter IV).

Quoting a small-scale experiment of my own with potatoes, using 50-yd. drills at 2 ft. apart, the following results were obtained. Compost and dung at 30 tons per acre.

	Per Drill *lb.*
Compost activated with 'Nitro Chalk' and super-phosphate	142
Ditto and artificial manure at 10 cwt. per acre	202
Dung only	142
Organic compost	190

The addition of 8 lb. of fertilizer costing about 1s. 3d. brought me in about 1s. 3d. worth of potatoes.

It will be noted also that the crop from the organic compost is about a third greater than that from the chemically activated.

A comparison of yields of pea plants in pots by Dr. E. Pfeiffer, of artificially manured and compost manured plants, suggested the following crop yields:

Compost, analysis N. 4 per cent, P_2O_5 0·5 per cent, K_2O 0·23 per cent, was added.

	Grammes per *Plant*
At the rate of 1 ton per acre, yield was at the rate of	14·4
,, ,, 5 ,, ,, ,, ,, ,, ,,	16·3
,, ,, 15 ,, ,, ,, ,, ,, ,,	16·1
Control	11·6

Using a fertilizer with an analysis of N. 6 per cent, P_2O_5 10 per cent, K_2O 4 per cent.

	Grammes per *Plant*
At the rate of 500 lb. per acre, the yield was	13·2
,, ,, 800 ,, ,, ,, ,,	13·0
,, ,, 1,000 ,, ,, ,, ,,	9·4

The last figure was less than the control.

A further experiment seemed to show that 55 lb. of nitrogen in compost is equivalent to 200 lb. in fertilizers. Considering the losses to which soluble nitrogen is subject this is not unlikely (Chapter IV).

Mr. Friend Sykes, of Chantry, in the dry season of 1949, achieved 60 bushels[1] of wheat, 68 bushels of oats and 72 bushels of barley per acre, and hay of 12–16 per cent protein content.[2] These crops are quite representative of the yields he obtains[3] on thin chalky down-land soil.

Mr. Newman Turner claims potato crops of 15–20[4] tons as an average and in a rather wet climate where blight is prevalent.

Lt.-Col. Sir J. E. H. Neville, Bart., of Sloley, Norwich, has kindly given me figures of his experimental plots in which he is running a comparative test between compost, farmyard manure and artificials.

Year	Crop	F.Y.M. and Fertilizer	Fertilizer	Compost
1949	Beet	11 tons	9 tons	8 tons
1950	Wheat	22 cwt.	15½ cwt.	22 cwt.
1951	Peas	14½ ,,	14½ ,,	19½ ,,
1952	Oats	22 ,,	15½ ,,	16½ ,,
1953	1 year ley	—	—	—

It will be noted how the first-season compost gave a comparatively poor yield, probably due to the low state of fertility of the ground. However, the succeeding crops show how this was quickly overcome, and beet now averages 11 tons per acre, with 16 per cent sugar content. His original reason for giving up artificial manures was because of heavy losses through lodging. These no longer occur.

Most of the farm of 550 acres is cropped with cereals of various kinds—wheat, oats, barley, and peas, which are sold off the farm. He also keeps 100 cattle, 200 pigs, 18 horses and 200 poultry, for which very little food is bought—and, with the exception of the horses, and poultry for eggs—are sold off the farm.

Turning to milk production:

Messrs. S. Mayall & Sons, Harmer Hill, Shrewsbury, have achieved an average for 1952–3 of 1,020 gallons per cow (Ayrshires), with a butterfat content of 3·95–3·98 per cent, after five years.

Lt.-Gen. H. R. S. Massey, Haverfordwest, Pembrokeshire, has had

---

[1] A bushel of wheat weighs about 63 lb.
[2] *J. Soil Ass.*, 4, 2, 11.
[3] Friend Sykes, *Humus and the Farmer* (Faber and Faber).
[4] F. Newman Turner, *Fertility Farming* (Faber and Faber).

a gradual rise in production from cows bought five years ago. These gave 5,000–6,000 lb. per lactation at first and have shown a steady increase to 7,000–8,000 lb.

Dr. E. Pfeiffer, in *Soil Fertility, Renewal and Preservation*, pp. 171–7, quotes increases of a similar order. In all cases increases were entirely due to a change-over to home-produced, organically-grown foods. Furthermore, sterility troubles have disappeared. Pfeiffer also quotes many instances of a gradual rise in cereal and other crops.

In many instances, large quantities of artificial manures have been used previous to conversion, and it may be considered that the crops are, in part at least, drawing on reserves of minerals provided by this means. However, Mr. R. Coward, near Shaftesbury, Dorset, is farming land that has had little or no artificial manures in living memory. He attains yields of between 22 to 24 cwt. of wheat, 22 to 33 cwt. of oats and 22 to 35 cwt. of barley; while peas crop at about 22 cwt. per acre. Swedes and hay are good, and potatoes at 6 to 8 tons are average for the county.

This is on a medium loam over greensand, which is very shallow in places. His cows average 8,000 lb., with 4·3 per cent butterfat. A large amount of produce is being sold off this farm from a soil of no great merit.

These are a random selection of figures from half a dozen farmers —there are now some hundreds exploiting the recuperative powers of natural processes. Once these methods are adopted by a farmer he does not wish to go back to the usual methods. Not only do yields generally improve, there is no longer the uncertainty of them and the fertilizer expense no longer occurs. The saving can be put to other purposes, and even if it does no more than compensate for extra expense in making compost where necessary, there is still the better crop and peace of mind. Furthermore, most of these farmers find that the plough is much less used, or perhaps not at all, and where it is, at a considerable saving in fuel owing to the shallower draught.

Take, for instance, the savings incurred in the sowing of a cereal after a root crop. Normal practice is to plough , harrow and roll, perhaps harrow and roll again unless the weather has been particularly favourable. On the easily worked soils of established organic farms a single rotary cultivation or discing and rolling may be sufficient, though two is more usual.

Let us see what this costs.

Using figures provided by the Public Relations Department of Harry Ferguson Ltd.:

Under orthodox methods:

Ploughing two 10-in. furrows at 6 in. deep	15s. per acre
Harrow springtime 6-ft. cut	3s. ,,
Roll, 6-ft. roller	3s. ,,
Harrow and roll again	6s. ,,
Total	27s. ,,

Under organic methods: One discing or rotary cultivation after the crop is cleared, leaving it until required for sowing for weeds to germinate and then cultivate again just before sowing.

Two discings or shallow rotary cultivations at
5s. per acre   10s. per acre.

This represents a saving on a 10-acre field of £8 10s.—well worth considering.

There are often other costs to be considered under orthodox methods. Preparation for a crop is often so hurried that no time is allowed for cleaning, and an infestation of weed occurs which at best will reduce the crop by a cost far in excess of that of an extra cultivation, and at worst will necessitate spraying at a cost of:

Materials	10s. per acre
Tractor spraying	1s. ,,
Total	11s. ,,

providing materials and water are on site, which they very often are not. The saving now amounts to £14 10s.

A further complication all too common these days is the lodging of cereals. Organic farmers seldom suffer from this. Good root systems and strong straw is usual. The usual cost of cutting is 4s. 6d. per acre. One-way cutting at least doubles this, giving an extra cost of 4s. 6d. per acre. The loss of crop can seldom be estimated, but even when it appears to be well saved the loss will certainly be greater than the extra cost of cutting. There is a risk then of losing nearly a further £8 on his 10-acre field even if he loses little crop. So far, organic methods would be worth risking even if the crop was smaller by 10 cwt. on his field.

A big saving is the lower seeding rate required. One and a half bushels of a cereal is sufficient instead of three, on account of the better germinating and growing conditions. An immediate saving of £40 on 10 acres.

The greatest saving of all is yet to come—the fertilizers. These will normally cost about £5 an acre, including lime—a saving of £50 on 10 acres. It may occasionally be necessary to make and spread compost, if only to dispose of yard muck, and this figure will be offset on these occasions. But, as will be explained later, compost, if used at all, is used on greedy crops like roots or kale.

The total savings then on 10 acres are as follows:

	£	s.	d.
On cultivations	8	10	0
Spraying	5	10	0
Losses due to lodging	8	0	0
Saving on seed	40	0	0
Saving on fertilizers	50	0	0
	£112	0	0

This saving is equivalent to over 75 cwt. of wheat at present prices (1953). The farmer can afford to lose this on the changeover without being out of pocket—he is more likely to gain this, and more.

There is no 'muck and mystery' about this, it is plain hard business —and those who get in first will reap the benefit.

In addition to these monetary savings, all organic farmers find that diseases of crops and stock fall steadily after conversion of the farm. Beans are seldom grown now, especially in the wetter districts, on account of chocolate spot. They are a particularly good food for animals and easy to grow. As far as my experience goes, the organic farm that cannot grow beans on account of this disease has yet to be found. Apart from lodging, rusts and mildews cause much loss. These again are seldom seen. While the degrees of sterility of animals so common these days are quick to disappear, and with the use of mixed leys, bloat is a rarity.

A further point worthy of consideration is that organically grown fodder can be fed in smaller quantities. It is more nutritious, claims being made of savings of 15 per cent in food bills. This may be due to a better balance of nutritive elements. When we see the analysis of fodder as '10 per cent crude protein', we do not know whether this is, in fact, protein or partly indigestible protein or other unusable form of nitrogen. In many clovers, for instance, only about half the crude protein is digestible protein. Or whether, if it is protein, the various amino-acids are in the same proportions as in a crop grown by organic methods.[1] They are very possibly not, and the animal may have to discard much of these amino-acids in order to get a sufficiency of those in short supply. It is often observed that a plant fed with much easily available nitrogen has a lower content of carbohydrate, which is a body-warming food.

Florida Experimental Station found that the greater the quantity of nitrogenous fertilizer added above a moderate level, the higher the percentage of nitrogen and water in the plant and the lower the dry weight[2] (carbohydrate). No total increase in weight resulted, so there was in effect a net decrease in dry matter with a higher level of fertilization. This sort of thing would occur with the high level of manuring used in market gardens. There are some grounds here for the belief that vegetables grown on artificial fertilizers are 'blown up with water'. A similar effect has been shown to occur in tomatoes grown in steamed soil.[3]

Rothamsted, too, in an experiment to increase the amount of protein in sugar beet leaves by spraying the leaves with nitrogenous fertilizer, found that it did so but at the same time depressed the sugar content slightly. The total crop was increased slightly also, so that the amount of sugar per acre remained the same—but there was probably more water with it.[4] The carbohydrate content of a crop grown under organic methods may be increased by the fact that they often have a longer growing season. Artificial nitrogen on, say, a cereal crop rapidly disappears so that the plant is hastened to maturity before making all the growth it might had the fertilizer been more steadily available.

A sample analysis of food grown on the organic section of the Haughley Experimental Farms is shown from the organic and mixed (manure and fertilizers) sections respectively.

AS PERCENTAGES

	SILAGE				HAY			
	Water	*Phosphate*	*Potash*	*Nitrogen*	*Water*	*Phosphate*	*Potash*	*Nitrogen*
Organic	69·7	0·097	·42	·65	12·2	·41	1·45	·92
Mixed	75·0	0·093	·35	·55	13·1	·26	1'00	·64

[1] For evidence of variation in protein composition through manuring see W. A. Albrecht, *Protein Deficiencies via Soil Deficiencies*, Part II. Reprint by Thurston Chemical Co. U.S.A.

[2] Florida Ag. Expt. Sta., 1949, *Bull.* 455.

[3] Jens Roll-Hansen, Rep. State Expt. Sta., Norway, 1952, 10.

[4] Rothamsted Expt. Sta. Rep., 1952, 67.

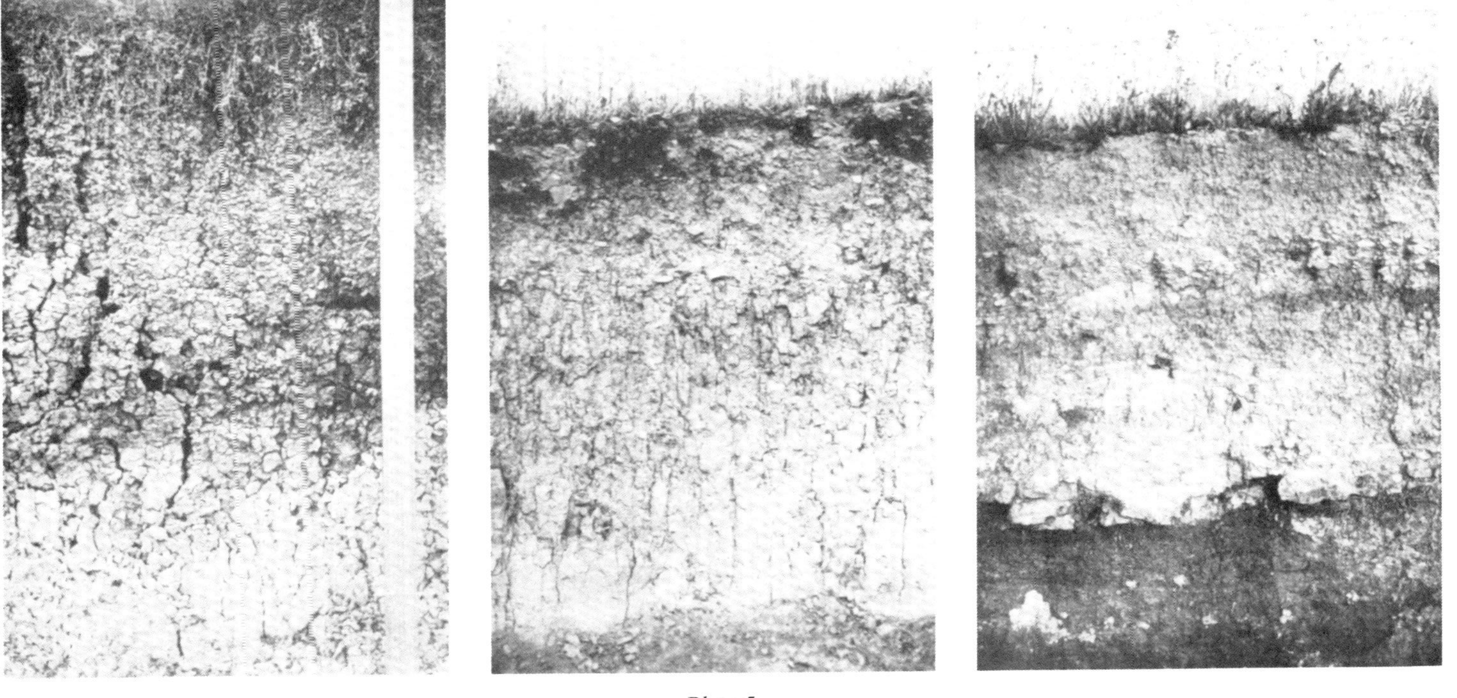

Plate 5

SOIL STRUCTURE

(*a*) Well-structured clay undisturbed by cultivation and well supplied with organic matter by the turf.
(*b*) Well-structured loam soil.
(*c*) Old arable land showing almost complete loss of structure and drainage. Topsoil is barely distinguishable from the subsoil.

(By permission of Mr. B. S. Furneaux)

PLANT ROOTS IN THE SOIL

*

(*a*) Showing vertical roots searching the subsoil for moisture and minerals and horizontal feeding roots. (Sweet clover).

Note that the largest plant (right centre) and another to the right of it have taken possession of worm-holes.

(*b*) Carrots. Taproot withdrawn from a worm-hole.

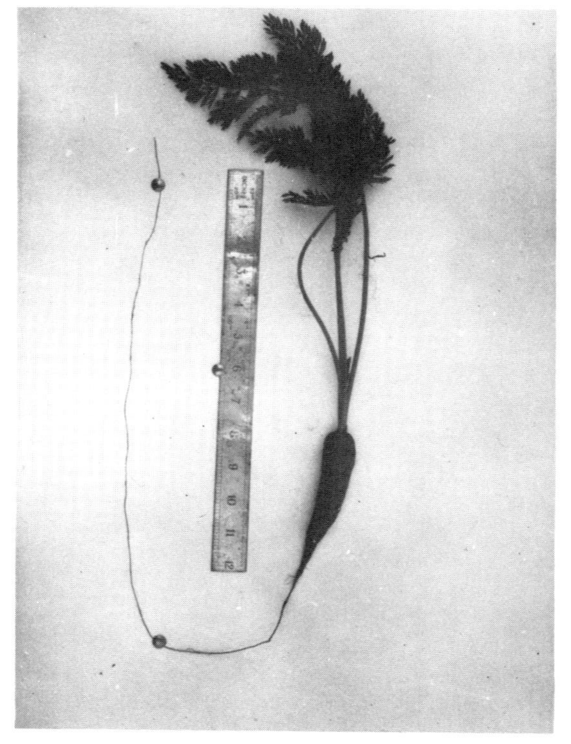

Solids in the silage represent 30·3 per cent in the organic and 25 per cent in the mixed, representing an increase in food value of about 20 per cent.

This is the general trend, though exceptions do occur.

This phenomenon of more nutritious food has also been recorded by Friend Sykes, Sir Bernard Greenwall, Dr. Pfeiffer, Lady Balfour and numerous others. The gradual decline of fats and solids in our milk over the last twenty years is quite probably associated with a decline in the value of foodstuffs owing to the vast increase in the use of artificial manures. Certainly Friend Sykes's experience of gradual improvement in the quality of his milk[1] tends to confirm this.

A few figures of yields of market garden and fruit crops will be of interest. There are a number available from my own holding, and though they mostly refer to small plots of a fraction of an acre, pathways have been included. It should be borne in mind also that these crops receive no top dressings, simply an application of compost at the appropriate season. Neither has any crop received artificial heat at any stage apart from an oil lamp at the propagating stage to keep out frost on cold spring nights. Figures and comments in brackets after each crop refer to previous yields under orthodox methods.

ANEMONES. 120 doz. bunches per 100 yds. cloches. August-March (variable, often complete failure).

APPLES. First year of bearing at 6 years old, 40 lb. per tree. (Trees are slow coming into bearing in wet districts.)

BEETROOT. 400 bunches per 100 yds. of cloches (not grown).

BLACKCURRANTS. 6-ft.-square plant, 3 tons per acre (not in cropping).

CARROTS. Sown mid-June between lettuce, 10–15 tons per acre (invariable failure through carrot fly).

CUCUMBERS. Cold house 200 sq. ft., 12 plants, four stems each, 4–500 saleable cucumbers. (Only for domestic use. Much red spider.)

LETTUCE. 600 per 100 yds. cloches. Sell at 50 per cent above prevailing rates. (Very variable, sometimes failed through mildew, slimy hearts or greenfly.)

LETTUCE. Outdoors, intercropped with carrots, cut about 30,000 from a plant of 35,000 per acre. Sell at 50 per cent above prevailing rates. (As above, often severe botrytis in addition.)

MARROWS. 800 per 100 yds. cloches. July-Sept., 120 plants (600 per 100 yds. cloches—much red spider).

---

[1] Friend Sykes, *Food, Farming, and the Future* (Faber and Faber).

ONIONS. 20 tons per acre, 4 ft. beds, 3 ft. paths (5 tons, usually failed to ripen, 'bullnecked').

POTATOES. 10–20 tons per acre (3–10 tons).

RUNNER BEANS. Double rows, 2 ft. apart started under cloches, 5 cwt. per 100 yds. cloches (3–4 cwt., much red spider).

STRAWBERRIES. Maiden plants, planted September, 3 ft. beds, 3 ft. paths, 8–10 tons per acre (4 tons).

TOMATOES. Bush variety, 2 ft. apart under cloches, 5 cwt. per 100 yds. of saleable fruit. Usually at least a further 2 cwt. wasted through unsuitable climatic conditions (i.e. slug damage, splitting). (2–5 cwt. per 100 yds. Much blotchy ripening.)

When comparing present with past yields it should be borne in mind that conditions in my district are very difficult for many of the crops grown as the early summer is cool and rather dry, while late summer and autumn are mild and wet, making harvesting difficult. The difficulties are greatly accentuated under orthodox methods.

In exceptional seasons even greater yields have been obtained—a bed of onions, for instance, in rather poor soil, was given an additional top dressing of compost at the end of June, which resulted in a crop equivalent to 35 tons per acre—averaging half a pound each. A strawberry plot, even though a quarter of the plants were removed owing to virus infection, cropped at the rate of 11 tons per acre.

Potatoes are not given any manure whatever, but are grown after sweet clover, as are tomatoes, whenever possible.

Other crops, cabbage, broccoli, parsnips and turnips, are grown for home consumption on too small a scale for figures to be given, but have always been satisfactory when given proper attention.

My annual bill for insecticides and fungicides for 2 acres actually under crops comes to about £2 for derris, copper oil emulsion and a proprietary insecticide for aphids in strawberries. In previous years I have spent £10; some of the stuff is still on the shelf and unlikely to be used.

Fertilizers for 2 acres of crops come to £2 for ½ cwt. of hoof and horn meal, plus three loads of dung at a cost of 30s., including my time and transport (dung is cheap locally), and about 10 cwt. of poultry manure at a cost of £5. In addition, some 5 tons of roadside trimmings are either dumped in for nothing or collected by myself, which can be given a nominal cost of 15s. a ton. This makes a total of about £12. This amount of 'fertilizer' is all that is used to produce £600 worth of produce.

I have, too, fertilizers left over from my first season, when £30 was spent, with a further £10 worth annually, plus 12 to 15 tons of dung and the roadside trimmings, making an annual expenditure of £20.

It is perhaps worth placing on record that I am the only market gardener, with no other source of income, left in the district—an indication both of the value of the methods and the deplorable economic situation of horticulture. It should perhaps be borne in mind also that much of my time is spent experimenting and observing which, though it may prove profitable ultimately, is quite unprofitable at the time. Neither have I the inclination or physical ability for sustained heavy work. Indeed, but for my change of methods, there is little doubt that I should have followed my neighbours. In actual fact, though most horticultural incomes declined steeply in the years 1949–51, at a time when the changeover was being made, mine remained constant, while in 1953, a bad year again for market gardeners, it rose by 20 per cent. By next year all the holding will have been built up by organic methods and further rises are confidently anticipated.

Mr. G. O. Liss, Hampton-in-Arden, Warwickshire, quotes[1] figures rising from 28 to 40 tons per acre for glasshouse tomatoes. The same house also produces chrysanthemums at over 100,000 per acre, and lettuce at about 50,000 per acre in the same season. This is after only four years of organic treatment on 'tomato sick' soil which also contained cysts of potato eelworm.

Mr. J. A. Bower, of Wisbech, records a number of interesting results since conversion to organic methods.[2] After taking over some naturally fertile land he found that crop yields began to decline under customary methods of artificials and dung. This became very noticeable after four years. His first success after practising organic methods was a 95 per cent cut of lettuce in a dry season when neighbouring crops failed. An experiment, using hoof and horn, bone meal and sulphate of potash after steaming, was made to compare the results from a simple application of compost made from the tomato haulms, also after steaming. The house suffered a bad attack of white fly and the plants under the usual treatment were destroyed. Those composted bore a good crop without 'feeding'. Where feeding with potash and nitrogen were attempted, in another trial, a slight reduction in yield occurred. Other comparative trials bore similar results.

---

[1] *The Grower*, 1952, 37, 2.
[2] J. A. Bower, *J. Soil Ass.*, Autumn 1947, 21; Winter 1947, 22.

A trial of lettuce on farmyard manure against compost worked out at 3,283 on composted land and 2,568 on manured land.

It will be useful to make a comparison of the growing of an expensive crop, such as tomatoes, under the different methods.

The average cost of growing an acre of tomatoes is as follows:[1]

	£
Labour	1,388
Fuel	685
Manures	150
Water	26
Seeds and plants	18
Sterilizing	59
Sundries	71
Overheads, maintenance, etc.	440
	£2,837

This gives an average yield of 35·5 tons per acre, costing 8s. 8d. per 12-lb. basket to produce, with a return of 10s. 8½d. The total returns, therefore, are about £3,532 per acre, giving a net profit of £695.

There are a number of items that will be eliminated or reduced under organic methods.

First of all, the labour.

	£	
No digging before sterilizing or to incorporate manure, say	50	saved
No labour for sterilizing	20	,,
Less spraying, say half	10	,,
No winter flooding	10	,,
Total	£90	saved on labour

---

[1] L. G. Bennet, Reading University. Report in *The Grower*, 14th November 1953.

```
No manure or fertilizers will be used—        £
    Manures and fertilizers at                150
    Instead, 30 tons of compost, not more than
        £2 a ton                              60
                                              ———
                                              £90  saved
                                              ———

    Plus saving on sterilizing (fuel and materials)  59  ,,
    Plus saving on labour above                       90  ,,
                                                      ———
                                    Total   £239 saved
                                                      ———
```

It must be borne in mind that the figures quoted for soil sterilization are an average for a number of nurseries, some of whom do not sterilize at all, some only occasionally, some with chemicals. For those who steam-sterilize their soils the savings are enormous—it costs between £300 and £500 per acre, according to various estimates. Some 40 tons of coal are commonly used which, at present prices, represents £250, and even under the most theoretically favourable circumstances it cannot be done with less than 20 tons. In addition, the growers who steam-sterilize annually spend more on water, spraying and fertilizers. So that the more progressive grower stands to gain much more than the backward.

These are the more obvious savings but, as shown already, crops of 40 tons are feasible (even with chrysanthemums and lettuce), giving a turnover of about £4,500, less costs at £2,360, leaving a profit of £1,870 against one of £695.

Where very efficient methods are already practised, much greater savings may be expected. In addition, there is the crop of chrysanthemums and lettuce, one at least of which cannot be taken under orthodox methods on account of the need for steaming, flooding and digging.

These figures are conservative—it seems that the experiment would be well worth risking on a portion of the nursery. It should be realized, however, that the salt concentration of many glasshouse soils is so high after many years of manuring that a digging, followed by prolonged winter flooding, will be necessary. And the compost will have to be worked in the top few inches until a good worm population is built up (worms may have to be introduced) which may take three years.

IV

PROVIDING THE FOOD

It remains now to discuss the mechanisms by which these crops are produced.

In order to produce good crops, plants need a steady and liberal release of nitrogenous compounds, potash, and phosphates, and all other necessary adjuncts to the plant's health.

All this is provided, under natural conditions, by one means, through the activity of soil organisms. These in turn require food, mainly as a source of energy. There is a great distinction here—the plant provides its own energy from light and often some to spare for symbiotic organisms. The micro-organism has to find its energy from ready-made materials but can often obtain its own mineral requirements direct from the soil particles.

The need of soil organisms for nitrogenous materials in the shape of dung, and so forth, has been stressed to the point of imbalance. Certainly they need nitrogen (and minerals), but they need fifty times as much energy-providing food. We give them ample roast beef but forget their bread and butter.

This energy supply consists of carbohydrates which are in turn plant remains, plant skeletons. It may be divided into various fractions varying slightly in their chemical constitution and considerably in their rate of availability as food to the micro-organisms. The most easily used fractions are sugar and starch; next come the hemi-celluloses and celluloses used for plant skeletons in herbaceous plants; and finally the lignins, found in trees and other hard, woody stems, sometimes to the extent of 30–40 per cent of the total carbohydrate. Lignin, it seems, can only be decomposed by fungi, whereas the first two groups can be utilized by bacteria as well. These materials are converted to carbon dioxide with the evolution of energy; indeed, the quantity of carbon dioxide liberated is a direct indication of the biological activity of the soil and consequently of its current fertility. (But not necessarily of its total or potential fertility, except under natural conditions, as biological activity can be stimulated by cultivations, addition of nitrogen and other nutrients, etc.)

A notable point of the natural cycle of plant growth and decay is its economy of nutrients. This is obvious enough in tropical climates where fallen leaves and branches decay rapidly and the plant foods they contain are equally rapidly assimilated by the surrounding growth. In most temperate climates, where most of the rain falls in autumn, the mildness and warmth of the soil at this time, coupled with the fact that plants can no longer grow very much on account of the shortage of light, decay and nitrification release a good deal of nitrogen into the soil which is in danger of being washed away. In fact, however, it is at this time of year that the fungi come into their own, feeding on the freshly-fallen leaves and stems, or dead grasses, and incorporating this free nitrogen into their protoplasm where it is held until the following year. Then, with a diminution in food supply and alternate droughts and rains of spring and summer, much of this fungal material dies and the nutrient is absorbed by the plants again. Worms also are active in autumn and assist by dragging plant remains into the soil or feeding on them, thus rendering them more suitable for fungal use. It would seem wise, then, to imitate Nature by applying undecomposed residues to the soil in autumn.

Bacteria, of course, are active more or less throughout the year, but their activities, which are greatest in autumn, would be rather in a secondary capacity after the plant remains had been incorporated with the soil by worms and fungi.

A soil well supplied with organic matter and in its natural uncultivated condition does not lose nitrogen readily by leaching with heavy rain for another reason, as explained on page 46. It appears that the water percolates rapidly down cracks and worm-holes to the subsoil, leaving the bulk of the plant food in the body of the soil granules or crumbs.

The ideal situation for this organic matter, it seems, is as near the surface as possible. Here again evolution can provide us with a clue. Plant and animal remains, with the exception of plant roots and soil animals, are deposited on the surface and dragged under bit by bit by worms and to a limited extent by other animals. Here it is either mixed up with finely-divided soil particles by the worm and excreted in an ideal form for further breakdown by bacteria and fungi, or it is acted on by these agents directly at the soil surface where the lower layers of litter are moist and shaded. It will be noted that the decay is invariably aerobic, encouraging nitrogen-fixing and nitrifying bacteria, and it is also adjusted to the plants' needs, giving an increased supply of plant foods in warm and damp weather when plants are growing quickly, and decreasing during cold or dry weather when plant foods cannot be used. A very different set of events takes place when organic matter is ploughed in, especially under modern conditions of deep ploughing and artificial manuring.

The plant and animal residues are put into a permanently damp atmosphere at the bottom of the furrow with a varying degree of aeration according to the depth of ploughing and type of soil. Providing the soil is not cold, e.g. winter time, decomposition is immediate and rapid, resulting in a large microbial and fungal population. Providing the residues contain $1\frac{1}{2}$ per cent or more of nitrogen, there will be sufficient of this element to support the plant as well as the micro-organisms. But also, if plenty of nitrogen is present, the carbohydrates in the organic matter will be used up quickly, giving a further and rapid release of nitrogen on the death of micro-organisms by starvation. The crop may be able to take up quickly all this nitrogen, but generally, as will be shown in a moment, excess nitrogen in the soil is liable to severe losses. This rapid evolution of nitrogen is followed subsequently by a comparative shortage and the plant may starve if not already mature, even though large amounts were added in the first place. When quickly available nitrogenous fertilizers or dung are used these effects occur even more rapidly.

There is another reason why the organic matter should be left on the surface. The carbohydrate of plant remains is converted to humus chiefly by fungi and to a comparatively small extent by bacteria. When these remains are turned under by the plough much carbon dioxide is evolved by bacterial action, which in turn suppresses the fungi,[1] so that carbon dioxide rather than humus is formed.

There are two, more serious, disadvantages to the use of nitrogenous manures not generally known—there appears to be no discrimination here between the inorganic forms and the organic forms of concentrated nitrogenous manure—such as dried blood, and hoof and horn. These difficulties again are associated only with present methods of agriculture.

It is obvious that a superabundance of a particular kind of food will result in a rapid increase in the particular kind of organism that enjoys that food: take, for example, the growth of yeast in a sugary

[1] N. R. Smith and H. Humfeld, *J. Agr. Res.*, 1930, 41, 97–123.

solution or the prolificacy of rabbits where there is plenty of winter grass. So it is when we add nitrogenous manures to the soil. These are added as, or are soon converted into, nitrates. Certain organisms can obtain both body-building materials and oxygen for their energy by breaking down nitrates—the denitrifying organisms. These can work without or with air,[1] but more commonly without—so the process is more rapid when the manures are turned under by the plough. They destroy the nitrates and convert them to atmospheric nitrogen—they are then lost to the plant. Indeed, the fact that these organisms can work under any conditions seems to indicate that nitrates have a short life in the soil. In warm weather this destruction of nitrogenous fertilizer takes two to three weeks under suitably moist conditions, that is why in market gardening top dressings have to be applied at about this interval.

Rothamsted show how an application of 14 tons per acre of dung (which contains a particularly high proportion of denitrifying organisms) to the acre loses 70 per cent of its nitrogen, while 4 cwt. of sulphate of ammonia annually lost 60 per cent. The higher the application the greater is the proportionate loss.

LOSS OF NITROGEN FROM A CULTIVATED SOIL

Broadbalk, Rothamsted—49 years, 1865–1914
As lb. per acre in top 9 in.

	Farmyard Manure Annually at 14 tons per acre	*No Manure*	*Complete Artificials, including 86 lb. N. as Sulphate of Ammonia Annually*
N. in soil in 1865—lb. per acre	4,850	2,960	3,390
per cent	0·196	0·114	0·123
N. in soil in 1914—lb. per acre	5,590	2,570	3,210
per cent	0·236	0·092	0·120
Total change in 49 years—lb.	+740	−390	−180
N. added in manure, seed and rain—per ann.	208	7	93
N. removed in crop—per ann.	50	17	46
N. retained or lost—per ann.	+15	−8	−4
N. unaccounted for—per ann.	143	(gain 2)	51

[1] J. Meiklejohn, 1940, *Ann. Appl. Biol.*, 27, 558–73.

This loss is equivalent to three crops of 30 bushels of wheat annually. Some of the loss may be due to leaching, but as this loss also occurs in dry climates it seems unlikely to account for much of it. It has also been suggested that the crop plant itself breaks down the nitrates—but there does not seem to be any need to postulate such a view.

In my own experience, plots of land receiving large quantities of dung and artificials for three years in succession gave no support to cabbage seed sown subsequently. While blackcurrant and other crops needing much nitrogen gave very satisfactory crops with a mere mulch of straw and no manure.

The best we can manage, when using concentrated nitrogenous nourishment for the plant, is to provide a rapidly fluctuating food supply, very dependent on rain for its solution and consequent assimilation, and with a risk of scorching the roots or depriving them of water in dry weather. It is for this reason that these crop yields vary so widely from year to year.

The other wasteful effect of nitrogenous manures not generally known is that it stimulates bacteria, including the nitrifying and de-nitrifying types, to search around for any possible source of carbonaceous matter for energy, and they will decompose the forms of humus (peaty material) which are normally fairly resistant to breakdown and have a great influence on soil structure. It is this effect that has, above all others, given artificial manures a bad name in many quarters—but it is not confined to artificial manures. We have the curious anomaly whereby the ploughing-in of a succulent green crop leaves the soil with rather less organic matter than before, owing to the fact that it contains much nitrogen[1],[2] (page 128). Even dung almost entirely decomposes itself, as can be seen in the case of a heap of manure left for twelve months on a paved yard. An annual application of 14 tons per acre is just sufficient to keep the organic matter at an economic level and, as already shown, the losses of nitrogen are enormous. If only for this reason, it is much more economic to compost dung either with fibrous plant residues or by spreading on grassland (sheet composting) so as to bring the proportion of nitrogen to about $1\frac{1}{2}$ per cent of the carbohydrate. Again, this effect is aggravated by our agricultural customs.

When ploughing-in a young green crop of, say, mustard or ryegrass, a rapid decomposition and release of plant foods takes place

[1] F. E. Broadbent, Proc. Soil Sci. Soc. Amer., 1948, 12, 246.
[2] H. D. Chapman and G. F. Liebig, ibid., 1947, 11, 388.

which unfortunately is completed in 1–2 months[1] and may be followed by starvation of the plant. So the plant has to be kept going with top dressings, hoping for rain to render them effective. Or it is given a heavy dressing of artificials, in the first place, in the hope that it will see the crop through—with a further loss of organic matter.

The above wastages take place when the dung or green crop is ploughed in immediately before spring sowing. It seems more usual, however, to plough in these materials in autumn. The losses of humus and nitrogenous plant foods are much more severe—the period of maximum nitrification coincides with the autumn rains and much of the soluble nutrient is washed away. We are faced again with the necessity of a spring top dressing.

Supposing, now, the crop of green manure is allowed to become mature and woody, and ploughed in. The proportion of nitrogen is much less than $1\frac{1}{2}$ per cent, probably more like $\frac{1}{2}$ per cent. All the available nitrogen in the soil is seized by the microbes to build up their bodies for the attack on the celluloses and their conversion to humus. The crop goes short. Another top dressing of fertilizers or basal application ploughed in, to compensate. There is, as before, a gradual decay and consequent release of plant food from this woody material. Unfortunately, the crop may not get the benefit from it until late in the season, or perhaps not at all. The decomposition may be completed by the onset of autumn and the wet weather will soon get rid of the hoped-for benefit. However, under the present system this is probably the best that can be achieved.

If, however, this woody material had been left on or near the surface, micro-organisms would not have been able to attack it in entirety, so absorbing all the soil nitrogen, as it would not have been in a permanently moist condition. It would also have provided the surface living nitrogen-fixing organisms with food to fix some nitrogen for the crop. Thus by ploughing in organic matter, not only do we not gain any nitrogen but we also lose some of what we had.

The losses of organic matter from the soil in spite of ploughing in green-manure crops are well demonstrated by the following table. In these experiments the green manure was sown in maize in the third week of August, and ploughed in during the first week of May before sowing the maize, over a period of five years.[2]

It will be noted that, generally speaking, the loss of humus is least where about a ton of dry matter, with a nitrogen content of between

[1] S. A. Waksman and F. G. Tenney, *Soil Sci.*, 1927, 24, 317.
[2] H. B. Sprague, N.J. Agr. Expt. Sta. *Bull.* No. 608, 1936.

1·2 and 1·7 is added. Crimson and alsike clovers, giving a large amount of dry matter but of a soft succulent nature high in nitrogen, cause the greatest loss. Winter vetch, on the other hand, forms a large weight of relatively mature tissue on account of the much better start in life provided by the relatively large seed reserve. Crop yields, not surprisingly, correspond with the amount of nitrogen added, but would decline over a long period where the soil organic matter was not maintained.

Plants Used	*Dry Weight of Plants lb. per acre annually*	*Nitrogen in Plants lb. per acre annually*	*Nitrogen Content of Plants per cent*	*Yields of Maize per cent*	*Loss of Humus as Carbon over 5 years, as per cent of Soil*	*Total Dry Matter Added over 5 years tons*
Weeds only	1,263	18·9	1·50	100·0	0·08	3·15
Winter Vetch	3,812	133·0	3·49	127·8	0·02	9·53
Crimson Clover	3,048	92·4	3·03	115·6	0·10	7·62
Red Clover	1,786	50·4	2·82	114·7	0·05	4·46
Sweet Clover	1,436	39·5	2·75	113·7	0·07	3·57
Alsike Clover	1,983	53·1	2·68	104·4	0·11	4·96
Winter Wheat	2,089	34·1	1·63	99·7	0·10	5·22
Winter Rye	2,463	31·8	1·29	95·3	0·10	6·15

It will be seen that the accustomed practice of repeatedly ploughing in young green crops and mowing grass orchards to 'build up fertility' is, in fact, doing the reverse—though in doing so it is stimulating the activity of micro-organisms in the soil, which will have some beneficial effects at the time. We must keep clear the distinction of 'building up fertility', i.e. nitrogenous humus compounds, which also preserve soil structure, and keeping up a high level of bacterial activity for some specific purpose, e.g. release of soil minerals. The two processes are largely antagonistic. One cannot save and spend until one has accumulated some capital.

It seems, then, that the more nitrogen used the faster the organic matter is used up. The less organic matter, the poorer the soil aeration; the poorer the aeration, the deeper one must plough and cultivate. The deeper one ploughs, the more nitrogen must be used in an attempt to balance the activity of the denitrifying organisms. The end result is soil bankruptcy and a hand-to-mouth existence for farmer and crop.

It should be mentioned that in some quarters it is considered that the addition of nitrogen to carbonaceous wastes conserves organic matter. The sort of experiment designed to show this uses carbonaceous matter with enormous quantities of nitrogen. For instance, Dr. F. E. Bear[1] added various organic materials of differing carbon-nitrogen ratios to soil at the rate of 2 tons per acre (4,000 lb. carbon to 2,000,000 lb. soil) as pure carbon. This means, probably, 4 or 5 tons of dry plant residues. To this quantity was added nitrogen to maintain the carbon-nitrogen ratio at 10:1, 5:1 and 1:1. Here are the results.

	Carbon-Nitrogen Ratio		
4,000 lb. of carbon added	10:1	5:1	1:1
Amount of carbon retained in soil	18,290	18,490	18,490
Carbon in control soil	16,360	16,360	16,360
Gain of carbon by soil	1,930	2,130	2,130

It will be seen that the two heaviest nitrogen additions resulted in less decomposition to the extent of 200 lb. of carbon, i.e. the greater the amount of nitrogen added the less organic matter was destroyed.

Let us examine these figures. The control soil contained about 0·8 per cent of organic matter. This is extremely low and is probably the very resistant residues from previous extractive farming methods, so that further losses from this soil would not be very great whatever its treatment. We may take it that the organic matter that was lost was from the added organic matter.

The quantities of nitrogen added are interesting. It was added as ammonium nitrate which contains 35 per cent of nitrogen. A normal dressing for a cereal crop would be 1½ to 2 cwt. per acre. To add 4,000 lb. of carbon to the soil with a nitrogen content of 10 per cent (i.e. C-N ratio 10:1) would require the addition of 400 lb. of nitrogen. The nitrogen in the original organic waste would not be more than 50 lb. if it were straw, so that the bulk of the nitrogen would come from the ammonium nitrate. This would need to be added at the rate of about 10 cwt. per acre which would have a slightly depressing or unbalancing effect on microbial activity on account of the salt concentration. To obtain a C-N ratio of 5:1 would require twice this

[1] *J. Soil and Water Cons.*, 1946, 1, 2, 83.

concentration, and one of $1:1$ ten times. It seems likely that the extra carbon remaining was due to depressed microbial activity—the microbes would be quite literally pickled. The enormous amounts used in any case bear no relation to farming practice. It would be interesting to repeat the experiment using commercial amounts of fertilizer giving a C-N ratio of around $30:1$ or even higher.

In view of these wasteful effects efforts are being made to manufacture a slow-acting 'synthetic organic' nitrogenous manure with some success. One such is formalized casein. We have already, in compost, a slow-acting nitrogenous manure, which not only does not reduce the soil's organic content, but actually increases it and makes for easier working of the soil. And it can be manufactured on every farm.

It has been said in many quarters (for instance C. Bould, *Sci. Hort.* 9, p. 28) that compost when used under normal fertile soil conditions begins to lose some of the 'halo' with which some advocates surround it. Though there is some truth in this, the 'normal fertile' soil conditions need qualifying. If by normal is meant a soil already well supplied with organic matter, then one would not expect much difference, as by adding compost one is merely maintaining the *status quo*. It is, however, suggested that the reason is that compost is low in available nitrogen, particularly if not fully rotted, and therefore extra nitrogenous manure is indispensable for optimum growth. If, however, the 'optimum growth' obtained by this method is simply extra water, the grower may gain apparently, but the consumer will not. When one considers the destructive effect on the soil organic matter of this extra nitrogen, and the waste of nitrogen involved, it seems probable that the farmer, too, has suffered a loss.

An experiment by Rothamsted appears, at first sight, to lend support to the view of low availability of nitrogen in compost. They are conducting an experiment[1] designed to find out the comparative values of various ways of using straw to keep up the soil organic matter. This has been in progress since 1934. The crops used were barley, sugar beet and potatoes in rotation, and the manurial applications were made every other year to test for residual effects.

Straw was applied at the rate of 53 cwt. per acre, with fertilizers, direct, or as a compost, rotted down. One series of plots was manured with artificials only. There were four treatments altogether. Fertilizers alone, as a control, straw ploughed in in autumn with fertilizers added in spring, straw ploughed in in autumn with half the fertilizers, the other half being added in spring, and straw compost alone.

---

[1] Rothamsted Expt. Sta. Rep., 1951, 135.

The barley had just sufficient balanced fertilizer to prevent nitrogen starvation, the sugar beet had more and the potatoes most of all.

The average yield per acre was as follows:

Treatment		Crop		
Autumn	*Spring*	*Barley (Grain)* cwt.	*Sugar Beet (Roots)* tons	*Potatoes* tons
None	N.P.K.	32·3	11·7	9·12
Straw	N.P.K.	30·8	10·9	9·64
Straw ½ N.P.K.	½ N.P.K.	30·8	10·9	9·25
Compost	None	27·5	9·9	8·00

Compost, it will be noted, gives the lowest yield of all treatments.

Rothamsted conclude that it may be dangerously misleading to regard straw as a basis for maintaining soil fertility, particularly as a compost.

Let us study the methods used in the experiment. First of all, the compost.

This was made by soaking straw in water, sprinkling with calcium cyanamide, rock phosphate and limestone. The amount of nitrogen added was 0·75 per cent of the weight of the straw, which, with that already in the straw, is considered just enough to rot it. It is the experience of many that organically activated compost, i.e. with animal or plant residues as a source of nitrogen, is superior to the chemically activated. As already shown, I have myself, in a trial, found that compost of dung and straw gave one-third extra crop of potatoes over that made with a similar amount of inorganic nitrogen with superphosphate and lime added, and about the same as this inorganic compost with additional potato manure at 10 cwt. per acre.

This compost was made in May and allowed to rot till November. To what extent it had rotted is not stated, but six months in warm weather is far too long to keep compost, without loss. But it does appear that it was exposed to the weather and suffered considerable leaching and decomposition, as the following figures show.

As cwt. per acre	N.	Phosphate	Potash
Straw and fertilizers supplied	0·7	0·5	1·1
Straw and fertilizers after composting	0·5	0·3	0·8

Nearly a third of the nitrogen, over a third of the phosphate, and over a quarter of the potash was lost.

Further, the compost was ploughed in during November when it would be subject to further decomposition and leaching during the winter.

The straw also was handled in such a way as to cause the maximum immobilization of soil nitrogen. The soil is too cold and wet in November to allow of appreciable decomposition of straw, and the nitrogen in the fertilizer added at that time in the half-and-half treatment would be subject to leaching and some denitrification. The remainder, added in spring, would be insufficient for the organisms rotting the straw, and the potatoes would go short. This would account for the smaller potato yield in this treatment compared to the fertilizers in spring treatment. The straw would commence to decompose about May and continue to immobilize plant nutrients until midsummer. This would also account for the drop in yield of barley and sugar beet where straw was used compared with artificials, both crops which are most susceptible to a check in growth during June when the fixation will be at its greatest.

Potatoes show an increase where straw and full fertilizer in spring is used; these are particularly appreciative of potash, which is provided to some extent by the straw which could be rotted in time for the potash to be utilized, by the full amount of fertilizer given in spring. Potash, unlike nitrogen, is not readily leached out of the soil (except sandy soils). Also, this crop makes rapid root and topgrowth during May and would seize a good deal of the easily available fertilizer before it was taken by the microbes rotting the straw. Particularly as the surface temperature is considerably higher at this time than that in the region of the straw, giving strong root activity.

Further evidence of a loss of nitrogen during winter on the 'half-and-half' plot is to be seen in the fact that although the barley yields are the same there is slightly more barley straw on the spring artificials plot (33·4 cwt. against 34·6), while in the case of sugar beet, though the weight of root was the same, there was more leaf and therefore more sugar in the spring artificials plot (9·0 tons of tops against 9·2 tons).

Sugar in sugar beet is formed when the plant is checked in its growth for some reason either by cold, drought, or lack of nourishment. It is probably for this reason that heavy dressings of nitrogen make sugar beet run to seed—a check in growth of a plant usually hastens development. It seems likely, then, that the smaller quantity

Plate 7

EFFECTS OF WORMS ON SOILS

(*a*) Showing the depth and numbers of worm-holes in a soil manured with green manure crops and surface cultivated.

(*b*) Top 2 inches of soil removed to show the density of worm-holes in a soil given 30 tons per acre of compost and crop remains to the surface annually for three years. (Pocket Knife is $3\frac{1}{2}$ inches long giving about 50 holes per square foot or 2 million to the acre.)

Plate 8

EFFECTS OF WORMS ON SOILS

*

(*a*) Showing how a surface crust formed by heavy rain after intensive rotary cultivation in dry weather can be lifted and cracked, thus aerating the soil in spite of adverse circumstances.

(*b*) Worm casts in old pasture during October when activity is high.

of nitrogen available on the half-and-half plot would rot the straw more slowly, and consequently release it again more slowly, also, later in the year, giving a steadier supply of nitrogen. On the other hand, the beet in the spring artificial plot would have an increasing amount of food as the straw rotted away quickly, followed by a sudden dimunition in food when rotting was complete, about August or September. Thus a large plant would be obtained, receiving a check, with all its resources turned to sugar formation.

That decomposition of the straw is much slower on the half-and-half plots is shown the following year when no further artificials or straw are used. Some plant food is still being released, as shown by the fact that all crops show the greatest yield over all other treatments in the case of the half-and-half plots. Fertilizers alone show a much more pronounced rise and fall between the manured and unmanured years than those receiving organic matter in the shape of straw or compost, showing that much less plant food is held over when there is little organic matter.

There is a gradual rise in productivity on the potato plots when straw or compost is used and it is attributed to the extra potash derived from the straw, which seems quite a reasonable explanation. Sugar beet have remained about the same over the period, except that the plot receiving artificials only has shown a decided rise over the remainder. Barley, however, has shown a progressive decline on all plots, especially in the case of the barley straw on the artificials-only plot.

It seems likely that the sugar beet and barley results are connected with loss of nitrogen and loss of organic matter simply through the nature of the experiment and the use of easily soluble nitrogen. It will be noted that barley follows potatoes—the crop receiving the greatest amount of fertilizer and cultivation. One consequence will be an increased destruction of organic matter by organisms living on the soluble nitrogen and therefore increased loss of nitrogen during the winter by leaching, on the death of these organisms, with further losses due to the cultivations. With the relatively small quantity of organic matter provided by a biennial application of 2 tons 13 cwt. of straw (15 tons of dung per acre per annum is just sufficient to maintain a reasonable level) one can expect a progressive decline throughout the experiment.

Sugar beet will have the manurial benefit of the stubble and weeds of the previous barley. These will prevent some surplus nitrogen from leaching away during the winter and in any case, being a deep-rooted

F

plant, it is well adapted to extracting its food from regions where organic matter is scarce (subsoil) and where some of the soluble nitrogen has been washed by rain. The fact that the artificials-only plots of sugar beet have an increased sugar content is very likely due to the already mentioned peculiarity of the beet plant—easily available nitrogen leads to rapid growth, followed by a check when the nitrogen is exhausted, and consequent sugar formation.

Rothamsted's view that it may be dangerously misleading to depend on straw as a source of fertility is quite understandable under the conditions of the experiment. When it is known that readily soluble forms of nitrogen are not only subject to severe losses in the soil but also stimulate soil organisms to destroy organic matter it is obvious that there never will be a rise in 'fertility' through using compost—it is destroyed as fast as it is added.

It is, it seems, this failure to recognize the nicety of balance between nitrogen, carbohydrate and humus in a naturally fertile soil that is the biggest stumbling-block of modern agricultural research.

I have been at length to suggest possible explanations of these results to show how, given a different viewpoint, i.e. of the importance of the soil micro-organisms, one can come to quite different conclusions.

This experiment, of course, to repeat one of Sir Albert Howard's complaints of research institutions, is not part of any farming system. There is no crop in the rotation to provide organic matter or fix nitrogen, e.g. grass or clover. The 53 cwt. of straw are puny beside the contribution of a ley, or dressing of dung, as providers of organic matter and improvers of soil texture. Therefore, the experiment is not readily applicable to ordinary farm conditions.

Straw, in any case, should have been turned in during August or September, while the soil is warm and before the autumn rains leach out the nitrogen. If this cannot be done on account of the cropping it should be composted in autumn and spread in spring, not the other way about, in order to obviate leaching. And certainly not left exposed to the elements. This, of course, leaves out entirely the constituents of compost: a farmer would be better off using some home-produced dung than bought fertilizers, apart from the more valuable product.

In support of this latter point, and in contrast to Rothamsted, an experiment lasting fifteen years at Sprowston, East Suffolk, was described by Mr. F. Rayns at a meeting called by the N.A.A.S.[1]

---

[1] Described in *E. Anglian Times*, 1st November 1951.

It was found that some form of nitrogen was necessary when adding straw to the soil to avoid the temporary locking-up of nitrogen. Using sulphate of ammonia, a 12 per cent increase over the control crop was gained. Folding sheep on straw spread on the land gave an increase of 21 per cent. Farmyard manure with the straw ('sheet composting') gave an increase of 33 per cent and was even more marked after the third rotation. This appears to be one of the very few deliberate experiments done showing the value of organic methods.

The value of organic matter is often quite overlooked in experimental work by our research institutions, as demonstrated by an experiment on the permanent barley plots at Woburn.[1] These crops have been grown successively for many years, with, more recently, the addition of nitro chalk as a spring dressing. Yields have become progressively lower until, in 1951, the crop was too poor and weedy to be worth harvesting and was ploughed in while still green. It was then decided to try autumn-sown barley against spring barley the following year. The crops were top dressed with nitro chalk the following spring. With the following results:

Cwt. of Grain per acre

	Winter Barley	*Spring Barley*
2 cwt. Nitro chalk	25·2	10·8
4 cwt. Nitro chalk	28·5	10·3
6 cwt. Nitro chalk	25·8	8·6

Reasons suggested for the success of the winter barley were that the crop was deep rooted or there were fewer weeds or to differences in soil acidity. The fact that a considerable amount of green manure in an ideal condition had been ploughed in a few months previously seems to have escaped notice. This would have completed decomposition when the autumn barley was sown and would have provided food for it during the winter and following early spring. By this time a plant large enough to take advantage of the nitrogen added as a top dressing would have developed.

By contrast there would be little food available from this organic matter for the spring-sown crop by the time it had developed sufficiently to absorb it. Neither would the plant be large enough to absorb much of the 'nitro chalk' which will be largely decomposed about three weeks after application by denitrifying organisms. The dry

---

[1] Rothamsted Expt. Sta. Rep., 1952, 152.

weather of early summer aggravated by spring cultivations on soil low in organic matter will ensure a poor supply of moisture. This latter point is borne out by the greater depression in yield with the higher levels of nitrogen—which would increase the concentration of the soil solution.

Here again is a perfectly feasible explanation for an orthodox problem, given a different approach. Such problems occur commonly throughout our experimental station reports, but almost invariably points of vital interest from the organic view are overlooked and omitted.

Again, experiments on forest-tree nurseries carried out by Dr. Crowther and others from Rothamsted showed that over five years of annual treatments with composts and fertilizers, in different combinations, results were obtained in the following declining order: Straw and bracken loaded with heavy dressings of fertilizers, composts made with fertilizers, heavy fertilizer, ordinary composts, light fertilizer, raw organic wastes with fertilizer, and unmanured. His conclusion is as follows: 'This series suggests that most of the observed differences can be ascribed to the nutrients in the compost and fertilizers with little additional benefit from the organic matter as such.'[1]

This statement does not agree well with the fact that straw and bracken loaded with heavy fertilizer and composts made with fertilizers were both better than heavy fertilizers alone, and that ordinary composts (bracken, hop waste, sulphate of ammonia) were better than light fertilizer. The readily available carbohydrates and organic matter have quite obviously assisted in the utilization of the fertilizers. The raw organic wastes mentioned are the unrotted materials, hop waste, chopped straw and bracken which would obviously immobilize a considerable amount of nutrient for some months until decayed. The results show even less agreement with those of their previously mentioned experiment on straw composts.

Dr. Crowther also makes the point that little if any of the nitrogen added to straw (1 part to 100 of straw), composted for two months at 30° C. (86° F.),[2] and then incubated in soil at 20° C. (68° F.), was liberated. This is to be expected. Sir A. Howard found that nitrogen was not liberated until three months had elapsed at very much higher average temperatures and, furthermore, with mixed materials (which normally decompose more rapidly than single kinds).[3]

[1] Rothamsted Expt. Sta. Rep., 1952, 45.
[2] Rothamsted Expt. Sta. Rep., 1948, 29.
[3] A. Howard and Y. Wad, *Waste Products of Agriculture*, 1931.

On the other hand, at Woburn, where experiments were started in 1942 to find substitute manures for farmyard manure for market-garden crops, the result of using composts with inorganic manures was clearly superior to inorganics alone. These were made with sewage sludge, farm wastes, straw and inorganic nitrogen, applied at the rates of 15 or 30 tons per acre annually. These composts were further fortified with phosphates and potash and inorganic fertilizers added as base dressings. In view of the liberal dressings of organic matter some build-up of organic matter should result, especially considering the exceedingly poor state of the ground at the start, and in fact the soil has been altered in texture to some extent. However, as quantities of nitrogen are used in addition, only the very resistant portions of organic matter will remain, containing only very slowly available fertility.

An indication of the extremely low state of natural fertility of Rothamsted soils and their dependence on soluble fertilisers is shown in the following table of the yields of a four-course rotation:

	Clover Hay cwt. per acre	Wheat bushels per acre	Swedes tons per acre	Barley bushels per acre
Unmanured	19	24	0·5	15
Mineral Fertilizers and Nitrogen given to Swedes	55	40	17	36

Swedes are a crop requiring relatively large quantities of potash which though present in ample amounts in Rothamsted soil is apparently unavailable through lack of microbial activity which in turn is due to a deficiency of organic matter. Lack of nitrogen, which swedes also need in fairly large amounts, is also evidence of low organic matter, such nitrogen as they are able to obtain is that absorbed by the clay subsoil, as shown by the fact that a slightly larger crop is achieved after fallow than clover. The clover, while providing some nitrogen from its roots, provided little organic matter as the tops were carried off, and only 15 cwt. per acre at that.

Fortunately, it is no longer necessary to take an indirect approach when discussing the ability of organic matter to render nutrients

[1] Agdell field, Rothamsted.

available—even where no inorganic nutrients have been added. There is now some direct evidence.

At the Haughley Research Farms,[1] experimental work has now shown that where a high level of organic matter is maintained, as on their organic farm section, as great or greater quantities of nutrients are rendered available as where inorganic minerals are added. This is evidenced both by soil analyses and the mineral content of the crop.

Take the following table of minerals removed from fields of barley bearing approximately the same crop in cwt. per acre.

Manuring	*Yield in cwt. per acre*	*Nitrogen*	*Phosphate*	*Potash*
Organic	18	0·33	0·12	0·07
Mineral Fertilizer	18½	0·28	0·11	0·08
Quantity of Mineral Fertilizer added	—	0·18	0·36	0·20

It will be noted, further, that where mineral fertilizers were used only a third of the phosphate was recovered in the crop and about two-fifths of the potash. This is quite a high recovery by orthodox standards and is evidence of good cropping management. This comparison of mineral uptake is representative of the crops as a whole.

Probably the most conclusive evidence of the release of minerals by microbiological activity is shown in the following graph, which shows the availability of minerals and nitrogen content as shown by soil analysis in the three sections of the Haughley Experimental Farms in Stockless (mineral fertilizers and green manures) Mixed (dung and fertilizers) and Organic.

It will be noted that in the Stockless section there is a slight rise in availability during the summer of potash, and a more pronounced rise in phosphate to about 25 milligrams per 100 grammes of soil— about two and a half times its winter level. On the organic section, potash and phosphate are at a similar level during winter, but rise rapidly as the crop grows in May and June to a level representing eight times as much by midsummer. This is followed by an equal swift decline with the ripening of the crop. It is interesting to note that the minerals are rendered available as the crop needs them and cease to be so as the crop ripens—a protection against possible loss.

[1] Haughley Research Farms Ltd, Progress Report, 1953.

1952 MONTHLY SOIL ANALYSIS

on a field of Barley in each Section of
Haughley Research Farms Ltd.—Soil Association Ltd.

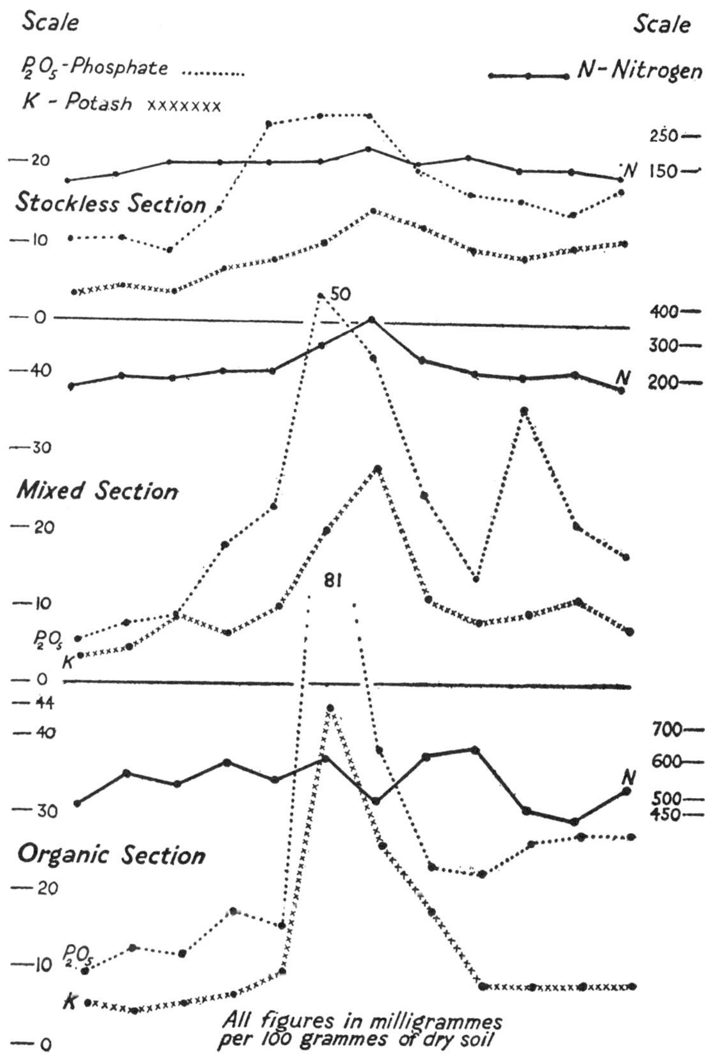

The mixed section shows an intermediate level. It is not inconceivable that some form of symbiosis occurs—the crop secreting food for the micro-organisms, while these in turn break down soil minerals for absorption by the plant.

There is another possible explanation. With the advent of warmer soil conditions breakdown of crop residues turned in during spring is hastened, especially under shallow cultivations, having the same effect on the soil organic matter as the Sudan grass in F. E. Broadbent's experiment, previously mentioned (page 74), giving a sudden but unsustained release of nutrients from the soil organic matter. This would be enhanced by cultivations exposing organic matter to oxidation. Nitrogen would also be released in large quantities, much of which would be lost through denitrification. (See also page 91.)

Taking this view of a high-nitrogen level as stimulating more rapid breakdown of soil organic matter, it is interesting to note that the highest figures of mineral availability are found in those fields with the highest nitrogen content. Further support is afforded by the fact that these peaks are least under leys—where there is no sudden addition of easily decomposable organic matter.

A third possibility is that the increased activity is merely a reflection of soil temperature giving increased biological activity where there is abundant organic matter, which is masked to some extent under leys, especially by day, by their insulating effect and the fact that no cultivations occur. The fall in July may be related to drought which was particularly noticeable in July 1952. On this basis a secondary rise when the ground became wetted in August or September would be expected, and though there is slight evidence of this it is not conclusive.

A fourth possibility is that the increase is largely a result of cultivations breaking up the soil and exposing more surface to oxidation which would take place as the soil warmed up. Almost certainly, all these factors play a part and their effect is enhanced by the abundant organic matter.

These possibilities can only be confirmed by intensive experiment which is at present beyond the scope of Haughley Research Farms who have opened up a whole new field of urgent research.

It has been noted, too, that the highest levels of available minerals have been found under barley crops. This is very probably a reflection of its comparatively low mineral requirements, the surplus remaining in the soil.

Providing the Food

A comparison of the levels of nitrogen, phosphate and potash in the three fields under barley will be of interest.[1]

Date	Organic			Mixed			Stockless		
	Available Phosphate	*Available Potash*	*Total Nitrogen*	*Available Phosphate*	*Available Potash*	*Total Nitrogen*	*Available Phosphate*	*Available Potash*	*Total Nitrogen*
January	10	6	500	6	4	210	11	4	150
June	81	44	580	50	20	330	26	10	200
July	38	26	500	42	28	400	26	14	240
November	27	8	450	20	11	240	13	9	170

The most noteworthy figure is the high level of total nitrogen in the Organic section—indicating a high level of Organic matter. The nitrogen level in the Stockless section did not rise by as much in June—when it was most needed—as that in the organic section, in spite of the addition of nitrogenous fertilizer, and, furthermore, the level of soil nitrogen over the whole farm is invariably lower where nitrogenous fertilizer is used. This is in line with the views put forward on pages 72–76. It is worth recording also that this organic field had grown four previous arable crops since being ploughed out of a ley five years previously.

This matter of the level of carbon-nitrogen compounds and their ratio is of some consequence to the organic farmer. Indeed, it may prove to be one of the focal points of research into the subject.

As mentioned already, these carbon-nitrogen compounds in the soil form a very large reserve of nutrient for plants—though generally in a slowly available form. They are derived from the decaying and decayed (humified) remains of plants and animals. When returned to the soil they have a carbon-nitrogen ratio of between 10:1 and 50:1, the former representing generally the soft nitrogenous plant remains and the undigested remains that have passed through animals. At the other end of the scale are the strawy, woody materials low in nitrogen. The end products of decay of these materials have a carbon-nitrogen ratio of about 10:1 (see Diagram). In order to provide a 'balanced ration' for soil organisms the

[1] Figures in mg. per 100gm. dried soil.

proportion of carbon to nitrogen should be about 30: 1, with the carbon in the form of carbohydrate—as is found in the newly-made compost heap and mixed natural vegetation. A consequence of this is that when material with a ratio above about 35: 1 (i.e. about 1·2 per cent or less of nitrogen to carbohydrate) is added to the soil the organisms have to draw on any free soil nitrogen —nitrates, ammonia, etc.—in order to balance their diet so that they can assimilate the carbon. This means, of course, that any crop grown on this soil will be starved of nitrogen and that the carbonaceous material will be slow to rot down. It is for this reason that such material should never be ploughed in but left on the surface, where it cannot compete with the crop roots.

DIAGRAMMATIC REPRESENTATION OF DECOMPOSITION OF ORGANIC MATTER IN SOIL

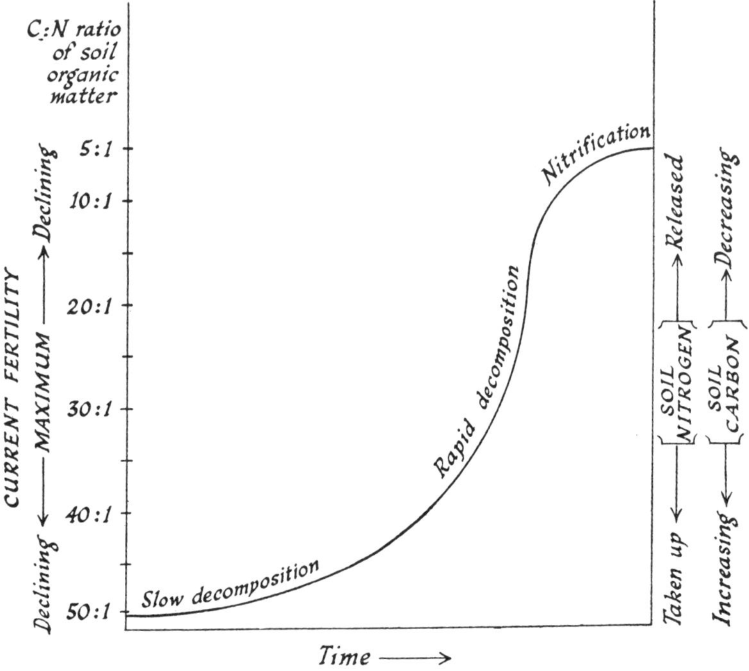

On the other hand, if material with a ratio of less than 25: 1, i.e. about 1·8 per cent or more of nitrogen to carbohydrate, is added to the soil there will be insufficient carbohydrate to balance the soil-organisms' diet. Two things may occur and it seems that both happen. The organisms may attack the more resistant materials of the soil humus in their search for carbon—thus destroying soil humus and releasing some further nitrogen, and creating a vicious though declining spiral. And, also, this surplus nitrogen becomes liable to wastage by denitrifying organisms and leaching.

The former is demonstrated by the following table.[1]

EFFECT OF ADDITIONS OF GREEN MATERIAL (SUDAN GRASS) IN INCREASING DECOMPOSITION OF SOIL HUMUS

Total CO_2 evolved	283 mg.	Soil alone
Part from Sudan Grass	225 ,,	19 mg.
Part from Soil	58 ,,	

The methods of the experiment need not concern us here but they seem to be incontrovertible. The amount of CO_2 (carbon dioxide) is a measure of the amount of carbon oxidized by soil organisms. The effect of adding the green manure was to triple the rate of loss of soil humus.

The amount of nitrogen released during this experiment was also considerable.

Total N. released	4·83 mg.	Soil alone
Part from Sudan Grass	1·93 ,,	1·4 mg.
Part from Soil	2·89 ,,	

The amount released from the soil by this nitrogenous green crop is half as much again as that in the added organic matter, and twice as much as is released by the original soil. The quantities released by the direct application of nitrogeneous manures without any carbonaceous matter whatever and the loss of humus and nitrogen involved does not appear to have been investigated. It is a very pressing need.

It might be expected at first sight that compost (with a ratio of about 10:1), and dung containing a high proportion of nitrogen, would have a similar effect. The carbon-nitrogen compounds in the former case are already at or approaching the condition of the soil humus compounds and its carbon and nitrogen is not very readily available to soil organisms. Dung is at an intermediate stage, having a good deal of its carbohydrate broken down into humus compounds by the animal, though still containing a considerable amount of the more resistant portions, including much of the celluloses and lignins,

[1] F. E. Broadbent, Proc. Soil Sci. Soc. Amer., 1948, 12, 246.

and also some freely available nitrogen. The addition of dung, then, will, if high in freely available nitrogen, as when fresh, result in a loss of soil humus and release of nitrogen, as in the case of the green crop but as it also contains some of these humus compounds, which the green crop does not, there will not be the same loss. If the soil was low originally in humus compounds there may be a gain. Take the following figures for a continuous wheat experiment.[1]

Additional Organic Matter Added per Annum	*Loss or Gain of Organic Matter* per cent	*Improvement in Organic Content over No Additions as lb. per acre over Twenty Years*
None	—0·24	——
Green manure of cow peas, 1 ton dry matter	—0·11	+2,600
Dung, 1 ton dry matter	+0·11	+7,000

The cow peas did not prevent a slight loss in organic matter compared with the original soil, but it lost 2,600 lb. less than the soil receiving no additions at all. The farmyard manure actually gave an increase in organic content over the original soil.

The inference to be drawn here is that in order to increase the organic content of the soil we must add carbohydrate material containing not more than about 1·8 per cent of nitrogen, or already humified material as compost but without extra nitrogen. Dung and immature green crops are relatively ineffective. Compost and mature green crops are.

As these humus compounds are not completely resistant to decomposition, especially when the soil is broken up and exposed to the atmosphere by ploughing, they will continue to break down slowly and the carbon oxidize away, releasing nitrogen for the crop. The final product has a ratio of about 4 : 1 and exists in quite small quantities. This is much the same ratio as bacterial protoplasm and these humus compounds may, in fact, be little else but remains of micro-organisms, having used up the last vestiges of food. Such material is commonly found several feet below the soil surface where little organic matter would penetrate.

A further result of the constancy of the ratio of carbon to nitrogen in the soil of about 10 : 1 is that it is impossible to raise the nitrogen content of the soil, except temporarily, without adding relatively

[1] C. A. Mooers, Tennessee Agric. Expt. Sta. *Bull.* 135, 1926.

large quantities of carbohydrate at the same time. The reason why the application of nitrogen to the soil is so wasteful becomes apparent. There is nothing in the soil with which the nitrogen surplus to the plant's immediate requirements can combine. It is, therefore, a readily available food for denitrifying organisms and those which derive their energy from the soil humus. Neither of these do we wish to encourage.

From the foregoing paragraphs it becomes clear why the few experiments done by our research stations on compost show little or no advantage from its use. Such experiments with compost are always done with additions of nitrogenous manures, and resulting crops are measured against those grown with similar total quantities of artificial nitrogen. The compost is, in fact, superfluous in such trials, and further, the contained nutrients, being more slowly available, the crops grown with it may not be as large as with a similar quantity of nitrogen as a soluble fertilizer. The use of nitrogenous manures in excess of the crop's immediate requirements also ensures that there is no build-up of humus matter in the soil, so that there is little improvement effected on soil conditions either. A fair trial with compost would require its use alone over a period of several years with minimum cultivations to permit a build-up of humus approaching that of an old pasture or woodland which, with special circumstances excepted, always yield good crops. The use of readily available nitrogen with compost is a biological contradiction. They cannot exist together.

It is commonly believed in orthodox circles that it would be impossible to farm on organic matter alone: it decomposes too slowly to release sufficient nitrogen for the crop. This, of course, is true enough on the great majority of our farm soils with an organic carbon content of about 1·2 per cent, with a low C.-N. ratio.

It is well known, however, that an old pasture, cleared forest or prairie, can grow abundant crops, in some cases for many years, without the addition of manures. Such soils have an organic carbon content of 2½ to 3 per cent and a C.-N. ratio of about 12: 1—i.e. they contain a high proportion of the less completely decomposed portions. They contain, in fact, material more readily decomposed and two or three times as much of it. The point to bear in mind is that the higher the level of organic matter the higher the rate of release of plant food.

The organic farmer, then, must try to build up a high level of carbon-nitrogen compounds in his soil with a fairly high C.-N. ratio.

DIAGRAMMATIC REPRESENTATION OF POTENTIAL NATURAL FERTILITY

Area enclosed is proportionate to the fertility.

The very high percentage of partially decomposed organic matter in the forest soil is probably due to an accumulation of litter in a cold dry climate.

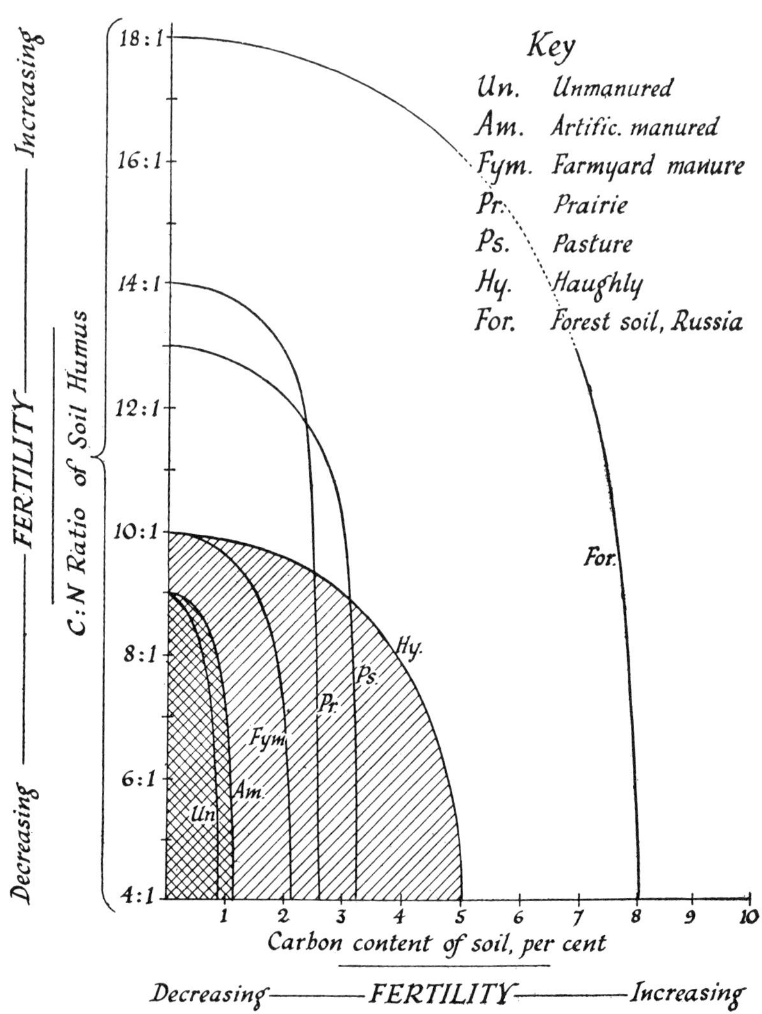

The simplest way of doing this is through the deep-rooted leguminous ley.

QUANTITIES OF ORGANIC MATTER AND C.-N. RATIOS OF SOILS

	C. *per cent*	*N.* *per cent*	*C.-N.*
Unmanured arable land, Rothamsted	0·89	0·099	9·0
Artificially manured arable land, Roth.	1·10	0·12	9·0
14 tons FYM annually, Rothamsted	2·23	0·22	10·0
Old pasture, Rothamsted	3·14	0·247	13·0
Prairie land, Brandon, Manitoba	2·48	0·187	14·0
Forest soil, Russia	8·01	0·450	18·0
Arable, Haughley, Stable Field (Organic matter Cx2)	5·0	0·5	10·0

Table showing the quantities of more easily available organic compounds (plant roots) as a proportion of the more resistant humus in virgin prairie soil.[1]

Depth	*0–6 in.*	*6–12 in.*	*1–2 ft.*	*2–3 ft.*	*3–4 ft.*
Weight of roots	2·6	0·72	0·63	0·30	0·04
Weight of organic matter	31·0	26·0	26·0	8·1	4·0
Weight of roots as percentage of organic matter	8·5	2·8	2·3	3·6	1·1

The C.-N. ratio is constant, more or less, for the type of farming adopted, the type of soil and the climate. Thus arable farming—which encourages oxidation of organic matter by constantly exposing it to the air, and returns little—has a low ratio through loss of carbon. A soil well supplied with lime also encourages a more rapid decomposition of organic matter than acid soil; so does a well-aerated soil compared with a waterlogged soil, and a warm climate compared with a cold one. These all tend to lower the ratio, all other things being equal, on account of their greater opportunities for biological activity. Cultivation apart, those factors which tend to lower the organic matter content are also good for plant growth: they encour-

[1] J. E. Weaver, V. H. Hougen and M. D. Weldon, *Bot. Gaz.*, 1935, 96, 389.

age nitrification of organic matter. Therefore, good soil and conditions need more organic matter—it is converted, indirectly perhaps, into plant food more quickly.

FACTORS AFFECTING DECOMPOSITION OF ORGANIC
MATTER IN THE SOIL

Increased Decomposition	*Decreased Decomposition*
Aeration	Waterlogging and compaction
High temperature	Low temperature
Much cultivation	Minimum cultivation
Alkalinity (i.e. high lime)	Acidity
High nitrogen and other nutrients	Low nitrogen
Moisture	Dryness

The destructive influence of cultivations on soil organic matter have already been mentioned—the loss of nitrogen can also be great. A measured loss of nitrogen from a prairie soil, high in original fertility, showed a loss of about 1 ton per acre in twenty-two years, exclusive of that taken up by the crop—which was only a third of this figure (Table, p. 48). Most of this occurred in the first few years of cultivation. This loss is due to the oxidation of the carbon on exposure to the air with release, and subsequent destruction by, denitrifying organisms of the nitrogen. It seems likely that much of this loss could have been avoided by shallow cultivations and undersowing the crop. The need for supplying the soil with quantities of organic matter could be very much less than is currently supposed.

The slowness of the rise in soil organic matter by the use of farmyard manure is well demonstrated by the tables derived from the continuous wheat field at Rothamsted[1] of the percentage of nitrogen in the soil which corresponds in quantity to the organic matter.

	Plot 2A *F.Y.M. from* *1885*	*Plot 2B* *F.Y.M. from* *1843*
1865	—	0·175
1881	—	0·184
1893	0·136	0·213
1914	0·191	0·251[1] (·236)
1936	0·186	0·226
1945	0·194	0·236

[1] Believed due to the inclusion of a piece of manure in one sample, the corresponding sample gave 0·236.

(*a*) EFFECT OF SOIL FLORA ON AGGREGATION OF
SOIL CRUMBS

Note that fungi are the most effective and bacteria the least.
(*By permission of Rothamsted Experimental Station*)

(*b*) THE EFFECT OF ADDING WORMS TO CLAY SUBSOIL

Left, dead worms added; right, live worms added.
(*By permission of the U.S. Soil Conservation Service*)

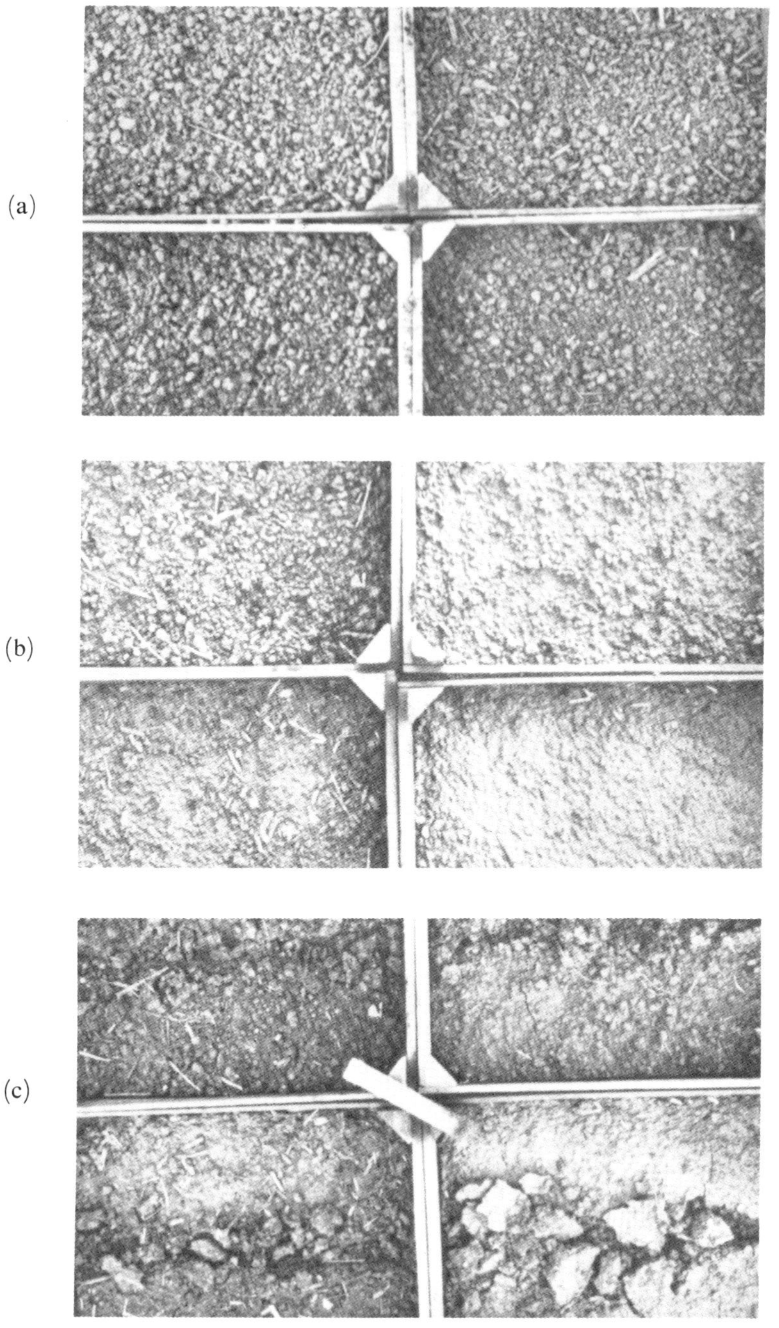

Plate 10

EFFECTS OF ORGANIC MATTER AND CULTIVATION
ON SOILS (I)

(caption facing)

Plots given no manure of any kind remained at about 0·10 per cent, while those given complete artificials keep around 0·12 per cent.

On the Haughley Experimental Farms the average nitrogen content is 0·25 per cent for the organic section, which has been achieved in about ten years simply through the use of leys, some compost, and elimination of soluble nitrogenous manures. One field, near the buildings, has a nitrogen content of 0·5 and organic carbon content of 5·0 per cent.

The presence of nitrogen as nitrates in the soil is not necessarily a good thing as these are liable to wastage. It is, in fact, a sign that there is little available carbohydrate in the soil. The two processes of nitrate production (either from nitrogen fixation or nitrification of organic matter) and nitrate absorption by other micro-organisms or plants go on side by side. In dry weather nitrate tends to accumulate, which shows a slackening in the rate of absorption or destruction relative to its formation, while in wet weather nitrate tends to disappear due to increased microbial activity, plant growth, leaching or activity of denitrifying organisms. With heavy rain, as in autumn, nitrates may be washed into the subsoil, or in light soils or wet districts washed out altogether. Ideally, then, there should always be a crop or source of carbohydrate in the soil to mop up this nitrogen. This is to some extent taken care of by the remains of crops after the crop has gone.

It will be interesting to attempt a calculation to find the quantity

---

Samples of soils under different treatments taken from the same soil type, (medium loam over limestone) within 100 yards of each other. Soil passed through a ½ inch sieve and exposed to approximately 6 inches of rain. A plant label was then drawn through the soil samples.

(*a*) Samples as taken:

Top left, 30 tons of compost per acre annually for the past three years. Surface cultivations only and no fertilizers used.

Top right, under grass three years. No fertilizers used.

Bottom left, 18 months under sweet clover 12 months previously. Surface cultivations only and no fertilizers used.

Bottom right, original soil, ploughed and fertilizers used, annually.

(*b*) After 6 inches of rain, showing complete disintegration of the tilth in the ploughed and artificially manured sample. Composted soil practically unaltered.

(*c*) Showing the type of crust formed. Note the hard lumpy crust under the usual farming methods and the soft crumbly crust under shallow cultivations and ample organic matter.

of organic matter required to maintain the biological activity needed to supply a crop with minerals. No experimental work has yet been done on the subject; we can only take an example from natural conditions. Ebermeyer has calculated that in a forest some 24 cwt. of carbon per acre are fixed from the atmosphere by photo-synthesis annually. About half this is returned to the soil as dead leaves and other litter each year. By a rough analogy, it would suffice, then, if half the carbon of our crops, which fix a similar amount or less, was returned to the soil. In actual fact, we seldom remove as much as half the plant when roots, stems and dead leaves are taken into consideration, and a much smaller proportion leaves the farm as produce or carbon-dioxide from animal metabolism. It may be considered that the minerals in a tree are in constant circulation, but in fact the wood retains about 4 per cent of minerals by weight, which compares similarly with the content of crops. Theoretically, then, about $\frac{1}{2}$ ton of carbon per year per acre should suffice for farm cropping if our present wasteful methods can be eliminated. This quantity would be provided by 4–5 tons of compost, while a good ley, providing about a ton of carbon in the roots alone, would be building up fertility.

Deep-rooted plants are most important for the organic farmer. If full advantage is to be taken of the soil we must use the whole depth of it—two, four or six feet. It is a matter of common observation that the only plants that thrive in poor and exhausted soils, provided there is depth, are the deep-rooted weeds—docks, bindweed, coltsfoot and the like. In pasture land the surface soil may be exhausted and surface-rooting grasses fail to grow, with a result that deeper-rooted species take control and the pasture becomes weedy. It is commonly stated that old pastures become very deficient in lime and phosphates, probably because that taken out in the animals' milk or bodies has not been replaced from below. Under such circumstances the weeds are probably the most valuable part of the sward, nutritionally. Unfortunately, they are not always palatable and perhaps even poisonous. It will be to our advantage, then, to sow a proportion of deep-rooted plants in our pasture and grow deep-rooted crops in our rotations in order to tap the considerable mineral reserves of the subsoil. Organic farmers will know of burnet, sanfoin, chicory, lupins, and others; while, for the market gardener, sweet clover is unrivalled, and lucerne, properly handled, for the fruit-grower.

Not only are such plants useful as providers of nutrients from normally untapped regions but their deep roots break up hard subsoils and, on decay, provide useful channels for drainage. Lucerne and sweet clover are especially useful in this respect in dry countries where irrigation is practised and periodic heavy rain occurs. In such climates there are no worms for this purpose.[1]

It is appropriate here to bring out a point which has puzzled me for many years, until recently. On the hedges and banks of this district (Pembrokeshire) there is a growth of 2 to 3 ft. sometimes 4 ft. of weeds and grass every summer. This is cut down and carted away at least once, often twice. Where do the plant nutrients come from? In the orthodox view this is an impossibility. It could be that hedgerow trees extract nutrients from surrounding fields and deposit them in their leaves—but often there are no trees or shrubs, just a bank or roadside verge. The answer, I think, is a superb justification of organic methods—no cultivations, a mixture of shallow and deep-rooted species maturing at different seasons. And, most important of all, sufficient herbage is permitted to grow late in the season to hold plant nutrient from leaching during the winter and provide food for micro-organisms—and ultimately the plant—the following spring. On this assumption the wise farmer would cut or graze his fields as much as possible until early autumn and not permit any further close grazing during the winter. It might prove economically sounder than allowing the ley to be eaten close in the winter to save hay.

It is often claimed by those advocating organic methods that irrigation or watering becomes unnecessary. Certainly with the increased quantity of organic material and improved structure, water-holding capacity is greatly improved, so that additional water becomes less essential and may often be dispensed with. As, however, most bacterial activity and consequent release of plant foods takes place in the top few inches of the soil, once this has dried out the plant's food supply is largely cut off. During summer weather, ground covered by a crop dries out to this depth in about a week under average conditions. Provided there is a depth of subsoil, holding moisture, the plant will continue to grow, though more slowly. In the case of market-garden crops and others, where the increase in value of the crop more than compensates for the cost of the water, additional water may be very beneficial.

The types of irrigation used are various but, provided the surface soil is moistened but not waterlogged and the surface is not compacted, the actual method is immaterial. The most effective system at present is the fine overhead spray, the finer the better.

There is one point to bear in mind when irrigating. The soil micro-

---

[1] Imp. Bur. Soil Sci. Tech. Comm., 1931, 22.

organisms will be breaking down organic matter at their maximum speed and it will therefore need to be replenished, under arable or market-garden conditions, more frequently. Grass orchards will be self-maintaining in this respect, the increased bacterial activity will be balanced by increased growth of the grass.

There is the problem of monoculture to be discussed. Again, from the evolutionary point of view, not many plants are adapted to growing in colonies of one kind. They may colonize the ground for a time, but soon give way to other species. Competition between members of the same species is the most intense as each individual has much the same power of acquisition of nutrients and much the same requirements. The difficulty is overcome in agriculture by giving each plant plenty of room to develop and, so far as we are able, the particular foods it requires.

It would be more economical of space to mix plants having different requirements, plants, for instance, that prefer sun and shade, occupy different root zones, have different mineral requirements or have different seasons of growth. Or even excrete materials which may be of benefit to one another.

This is commonly done in the case of the mixed ley and the growing of mixed corn for feeding animals whereby an extra 2 cwt. of hay[1] or corn[2] may be obtained per acre. The market gardener, too, has customary mixtures—lettuce and carrots being a popular one. Unfortunately, for economic reasons, mixed market-garden crops are unpopular, it being found generally that the extra labour involved is not compensated by the extra crop.

On the other hand, some plants have a depressing influence on one another. Grasses and cereals are powerful extractors of nitrogen and check the growth of neighbouring plants. Couch grass has a depressing effect on the germination of mustard seed, apparently due to some secretion.

There is certainly scope for much research into combinations of crops, particularly economically practicable ones.

In this connection, too, of course, diseases and pests have a vastly increased opportunity to multiply where one species is grown, and intercropping might, in some instances, be beneficial for this reason.

When one considers the enormous quantity of soil organisms in relation to the weight of farm animals foraging on the surface an

[1], [2] For many references, see E. J. and E. W. Russell, *Soil Conditions and Plant Growth*, 1950, 483.

interesting point arises. With estimates of soil organisms as weighing between 10 and 40 tons per acre, there is a proportion of mammals to microbes in the farming system of 30:1 up to 120:1. Even under the high mammalian population of modern agriculture the proportion of dung they provide is very small against that of the soil organisms. The high evaluation of mammalian dung has arisen probably because it was the only form of dung that could be seen and handled and because of its undoubted effect on plant growth. The fact that there is about fifty times as much excreta from soil animals and microbes is liable to be forgotten. Which puts a highly controversial proposition to the organic world. Is dung of warm-blooded animals really necessary? Can an agriculture be pursued with very little or even no mammalian dung? When one considers that flowering plants occurred before the existence of warm-blooded creatures and their comparative paucity under natural conditions, it seems likely we can, using a plant fixing much nitrogen and carbohydrate in the rotation. There are still the droppings of birds, mice and many other small animals, for what they are worth. And, as we have seen before, farmyard manure applied to bare land is very wasteful of nutrients and at best an ill-balanced manure.

It is not suggested that farmyard manure is of little value. If it is on the farm, use it by all means, but in small, intensive horticultural units it is not always easy and sometimes economically impossible to fit in stock without resorting to bought feeding stuffs. And, in such cases, it is often cheaper to buy dung direct, for composting, rather than the feeding stuffs.

There is further considered to be some special virtue in dung in some quarters over and above the food that the animal consumes.

These virtues are considered to be in the shape of trace elements, hormones and vitamins. Trace elements are not created by the animal. Whether mammalian hormones are of any value to plants generally is debatable. Hormones in plants are of a specific nature and purpose and found in particular parts of the plants. It seems unlikely that, except in special instances, a heterogeneous mixture of hormones would cause anything else but confusion when applied to plant roots. There is evidence that extracts of farmyard manure may increase root formation. Whether this is necessarily of value to the plant is again debatable. The plant is well adapted to managing its own affairs, and in any case such substances appear to have a very short life in the soil.

As regards vitamins, so far as we know, those found in the soil are created by soil organisms and plants equally well from dead plant remains direct.

It is considered a good thing, too, on a market garden to keep a pig to consume waste vegetables to make dung. There is, in fact, no more nutrient in the dung than in the food that the animal ate. Rather less. Certainly the proportion of nitrogen and minerals will be higher in the dung because much of the carbohydrate in the food will have been broken down by the animal, and the nutrients are therefore more readily available to the plant. It is, in fact, rather poorer in energy providers and in this respect poor food for micro-organisms. While not denying the possibility that dung may have some especial virtue, it is very likely that the soil microbes would be even more appreciative of the barley meal applied direct than after the pig has had his share. The basic fact remains that the wholesale use of the dung of warm-blooded animals finds no counterpart in Nature.

Once one had rid oneself of the complex of ideas associated with present methods, and grasped the fundamental technique of feeding the soil micro-organisms, organic agriculture is the simplest and most foolproof method of farming ever devised. Furthermore, while it is a relatively simple matter under orthodox methods to make two blades of grass grow where one grew before, organic methods provide a means of maintaining this economically.

Before going any further, it will be well to get one point clear—what is fertility? A debatable point, but it might fairly be described as the capacity of the soil to produce crops and go on producing them.

In the orthodox view, this consists of providing the plant with plenty of soluble nutrients. We have seen the wastage that occurs and the erratic and rapidly declining rate of food supply for the plant, with loss of organic matter and soil condition.

In the organic view, fertility consists of maintaining a high level of carbon-nitrogen compounds in the soil, of widely varying degrees of availability, which decompose gradually to give the plant a slow or rapid rate of food supply as the plant requires it and maintain a high degree of biological activity in the soil.

These compounds comprise plant and animal wastes, their decaying and decayed remains, and, most of all, the bodies of soil organisms as living protoplasm—unaffected by the leaching of rain and other wastage. While the organisms themselves extract the more insoluble minerals from the soil.

To maintain this high level we must add as much plant waste (carbohydrate) to the soil surface as possible, with nitrogenous compounds, preferably in the region of 1–2 per cent of nitrogen to carbohydrate. To conserve these carbon-nitrogen compounds we must make the minimum of cultivations and use the minimum of quickly available nitrogenous manures. From the figures of losses given, it seems likely that the apparent need for organic matter and nitrogen could be cut by a half or three-quarters.

This high level of carbon-nitrogen compounds is the key to organic farming and cannot be stressed too strongly. These compounds are plant food, soil aerators and water-holders rolled into one—and the only possible way to conserve nitrogen in the soil.

One may look upon the soil as gilt-edged stock—expensive of capital, but the dividend is safe. The alternative is a gamble at best— and gambling is expensive in the long run.

<h1 style="text-align:center">V</h1>

THE PROBLEMS OF PRESENT METHODS OF AGRICULTURE

From the time of Boussingault and Liebig it has been assumed that what the plant has extracted from the soil as shown by analysis must be returned if fertility is to be maintained. It was thought at first that the return of these minerals in the quantity extracted was all that was required. It was then found that much larger amounts were required, and in a soluble form. The reason for this is now well known: these soluble potash and phosphate salts are quickly converted into insoluble compounds by chemical combination with other minerals in the soil. 'Fixation', as it is called. This takes place as soon as they are washed into the soil—in 1–3 weeks. It is said that a plant takes up most of its phosphate in the first few weeks of life anyway—but it seems that the plant has no alternative. The apple-grower's dilemma of a soil rich in potash while his trees starved is a classic. The cure was simple—the soil was sown to grass and mowed. The soil micro-organisms fed on grass trimmings, released the potash, and the situation was relieved.

The compounds formed in the soil vary in their degrees of availability according to the type of soil. Taking the present-day view, in those soils where there is a very strong fixation of minerals it is difficult to imagine how anything ever grew before men tilled the soil, but there is little doubt that in many cases it grew forest giants whose appetite was considerable. The argument offered, of course, is that the little available is kept in circulation and, once taken away, more must be added. The fact that most of our land surfaces were many thousands of feet higher than at present, and that this amount of rock while being dissolved away by natural agencies must have released, and lost, colossal quantities of plant nutrients is overlooked. Under natural conditions it is estimated that 50 tons per square mile of

104

soluble inorganic materials are carried away in the river water every year in temperate climates, i.e. about 1¾ cwt. per acre.[1] Most of this is calcium carbonate, but there are appreciable quantities of magnesium and potash removed also, together with many other elements.

Furthermore, potash and phosphate have an almost unique peculiarity in that the bulk of them remains behind in the soil after the remainder of the rock constituents have been dissolved away. River water contains a much smaller proportion of them than the rock over which it flows. Indeed, it is fortunate that they are not readily soluble —or there would soon be little left for the plant. The reason for this is that these two materials rapidly recombine with other constituents in the soil—potash with the aluminium compounds in clay, and phosphates with aluminium, iron and calcium. It is probably this abundance of potash and paucity of sodium which is readily soluble that necessitated the plants' evolution from the sodium metabolism of its marine ancestors to its present-day dependence on potassium. It is noteworthy that plants of maritime origin can still use, or have re-developed a capacity for using, sodium in quite large amounts.

It is true that geological formations differ in their rate of erosion and, therefore, the release of plant nutrients. Those which are slow to break down generally form the thin stony soils of our high land, and would by their nature be easily recognized as poor. On the other hand, in some tropical countries where microbial activity is intense, amounts of nearly 1,000 lb. of potash per annum can be extracted from soils overlying potash-rich rock.[2]

The fact that one has to add very much more of minerals as fertilizers to the soil than one can extract, presents us with a paradox. The original stocks of minerals in circulation would soon have vanished to nothing if even a small proportion was fixed each time it passed through the soil.

The actual quantity of superphosphate taken up by the current crop would shock most farmers—it is 2 per cent. With a possible further 3 per cent in succeeding crops. This covers phosphatic fertilizers generally. Only 20–30 per cent of the phosphate content is recovered under present methods.

We go to the trouble of importing rock phosphate, and sulphur to make sulphuric acid with which to treat the rock, at the cost of large amounts of labour, machinery and coal. Only to have this soluble superphosphate return to often even more insoluble forms than the

---

[1] Data of Geochemistry, F. W. Clarke.
[2] A. S. Ayres, Proc. Soil Sci. Soc. Amer., 1947, 11, 175.

original rock in a matter of a week or two. The wasteful futility of this procedure must be apparent to our soil scientists and manufacturers.

It is interesting to learn that under an experimental scheme in Somerset, New Jersey, U.S.A. farmers are being paid eight dollars a ton to use rock phosphate. Results are said to be good. There is plenty of evidence that the mineral has given good results under organic methods.

During the war Rothamsted experimented with mineral phosphate rock and found it gave good results, especially for swedes on acid soils in wet districts. F. C. Gerretsen, of Holland, speaking at the 4th International Congress on Microbiology 1947, gave evidence of the greater utilization of inorganic forms of phosphate by soils containing a good supply of organic matter. Many strains of bacteria have been isolated which can extract these phosphates under such conditions.

Potash is fixed in very variable amounts—from 20 to 80 per cent may be recovered in the first crop and a total of 30 to 100 per cent in further crops. The amount recovered depending on the compounds formed in the soil and the organic content of the soil. Too much potash upsets the balance of other minerals, making it difficult for the plant to extract them. The magnesium deficiency of tomato plants in glasshouse soils is a well-known example. It is now claimed that one should analyse one's soil annually and use fertilizers with discretion, but only since these facts were recognized. It is generally realized, these days, amongst the more enlightened and progressive authorities on soil science that organic matter plays a great part in rendering minerals available. But it is far from being generally accepted that they can be released in sufficient quantities to maintain our present standard of cropping. Many, however, believe that it can be done on most soils, and while it is highly desirable that urban wastes be returned to the soil, there are in fact abundant reserves already there (see table on page 30). Those soils proved to be very deficient in, say, (total) phosphates should not be used for agriculture (except forestry) except under desperate circumstances, and here again the rapid fixation of phosphates suggests that there is little advantage to be gained from adding soluble phosphate, except perhaps in the initial stages of reclamation.

The effect of sulphates in the soil has received some attention. Calcium sulphate is not normally found near the surface of natural soils, being slightly soluble in water. It is, in fact, the responsible agent for permanent hardness in our water supplies. It is, however, formed in surface soils in equivalent quantities to the amount of sulphate of ammonia, potash or superphosphate added to the soil. The latter compound, as a matter of fact, already contains about 60 per cent of calcium sulphate. In poorly aerated acid soils particularly, in which the organisms responsible for the breakdown of organic matter cannot thrive, and in those soils badly aerated through mismanagement, including a deficiency of organic matter, these sulphates are broken down by bacteria, giving hydrogen sulphide—a substance very toxic to plant roots and most soil organisms. Thus a vicious circle is set up and organic matter continues to accumulate, while the agents of its decay are inhibited from doing their job of aerating the soil and releasing plant nutrients.

Recent work[1] has shown that in soils easily rendered acid, a species of bacteria has been shown to produce hydrogen sulphide from sulphates even under aerated conditions. This only occurred, however, in the absence of fungi, and although no antagonism between this bacteria and the fungi was shown, the fungi appeared to inhibit the production of hydrogen sulphide by the bacteria. The practical application of the work would be that in soils incongenial to fungi, either through depletion of organic matter, sterilization, waterlogging or acidity, the use of artificial manures in the form of sulphates would be dangerous. And, conversely, those soils rendered poisonous to plants by the presence of hydrogen sulphide can be improved by adding organic matter and draining.

This type of thing may occur naturally on peaty heathland, the classic example being Wareham Heath in Dorset. Dr. M. C. Rayner found, in her endeavour to make forest trees grow on this site, that the vicious circle of toxic soil—no mycorrhizal fungi, no trees—could be reversed by the addition of composted organic matter with its complex microbiological content. Incidentally, the completely organic compost was more effective than that activated with artificial nitrogen, while that activated with ammonium sulphate was mediocre. Not surprisingly, if sulphates were partly the cause of the trouble. It is significant, also, that those composts made of easily available carbo-hydrates, e.g. waste hops, were better than those made from sawdust—with straw taking an intermediate position. One compost made with ammonium phosphate gave very nearly as good results as the best organic compost. As both these contained considerable amounts of phosphate, it seems that this was a direct manurial effect,

[1] Rothamsted Expt. Sta. Rep., 1951, 57, 1952, 60.

but the organic compost maintained growth slightly better in the second and third years.

It was also found that a complete fertilizer mixture gave good results on this soil.[1] In this case, of course, the tree would be receiving nutrients direct instead of through the mycorrhiza. However, annual dressings of fertilizers seem pointless when a single dressing of compost is all that is required.

At the Woburn experimental farm (Rothamsted) a plot of ground received sulphate of ammonia for fifty years prior to 1926, when it had obviously become sterile through acidity. No lime has been added and the plots have remained sterile ever since, apart from one or two acid-tolerant grasses. Thus, when a farmer counts the cost of his artificial manures he should include the cost of the liming to counteract it as a direct charge. It is this loss of lime that leads to the curious anomaly of limestone or chalk soils being in need of lime.

Further evidence of a reduction in the microbiological activity of the soil by the presence of artificial manures, perhaps connected with the presence of sulphates, is shown by experiments made at the Haughley Research Farms. A wad of cotton wool was rotted away to the extent of 83 per cent after having been left in virgin soil (rough grass) for six months, while similar soil treated with complete fertilizer gave only 48 per cent decomposition. The virgin soil, with organic compost added, rotted the cotton wool completely in the same time. It is only fair to add that Rothamsted experiments, using a field permanently cropped to mangolds, have found consistently that the unmanured plot gave the lowest numbers of cellulose decomposing microbes, the artificial plot an intermediate number, while the farmyard manure plot gave the highest.

It seems that the rate of decomposition is in some way connected with the quantity of organic matter present, particularly the easily decomposable portions. Rothamsted soil plots receiving no manures would have rather less decomposable crop residues than those receiving artificials on account of the smaller crops, and much less than those receiving farmyard manure. That receiving farmyard manure would contain the most organic matter. On the other hand, the Haughley soil which had artificials added and was originally high in organic matter would have this content reduced somewhat by the presence of easily available nitrogen, while the addition of compost would increase it. Possibly the cellulose decomposing microbes require some material provided by the organic matter. In any case,

[1] Rothamsted Rep., 1946, 31.

numbers of bacteria are not necessarily an indication of their activity. Their rate of metabolism must also be taken into account. This may be retarded by the presence of inimical substances.

Long Ashton Research Station recently announced[1] that calcium sulphate (gypsum) aggravated 'whiptail' in cauliflowers. This complaint is also due to a deficiency of molybdenum and worsened by acid conditions. It could be rectified by adding lime at 6 tons per acre, or or sodium molybdate at 4 lb. per acre. An example of trace-element deficiency probably aggravated by the use of artificial manures with their high sulphate content. These plants also contained two to three times as much nitrate as healthy plants—not very good for the consumers.

It is known that sulphates are toxic to plant roots—a solution of 1 in 2,000 of sulphate of ammonia has a repressive influence on tomatoes, for instance—and the concentration in glasshouse soils often exceeds this by a good deal, especially locally, after a top dressing has been watered in. The same effect occurs outdoors if the weather turns dry after sowing the manure. This will render the plant susceptible to drought in two ways—by checking root development and, by creating a stronger soil solution, make it more difficult for roots to extract moisture. This, in addition to the ploughing pan, is expecting too much from any plant. Hence the need for rain after a week of fine weather.

Sulphate of ammonia and other soluble fertilizers have a detrimental effect on earth-worms. An experiment carried out at the Haughley Research Farms,[2] in a compartmented box, with free access to and between compartments. The box had six compartments and five worms were added to each. The soil in each compartment was given a different treatment with the normal quantities of fertilizers and composts used commercially.

After three days the worms had gone from the compartment containing soil with sulphate of ammonia, while they did not all leave the control soil until six days had passed. For some reason, however, a compartment treated with a complete fertilizer kept one worm for a fortnight—the length of the experiment—though the rest left within a few days. The majority of these worms accumulated in the soils containing compost, particularly the one with 'Indore' compost, and to a lesser extent a compost made with sawdust, and one chemically activated. The basic need for these creatures seemed to be food in the shape of organic matter, but it is evident that the presence of arti-

---

[1] W. Plant, Long Ashton Report, 1950, 91.
[2] *Journal of Soil Ass.*, Harvest No., 1947.

ficial manures was not to their liking. It is difficult to see why one lingered so long in the complete NPK treated soil. Possibly it was so overcome by the combination of artificials that it had not the strength to move.

It is true, however, that this experiment did not represent field conditions where artificials are usually worked into the top 3–4 in. The worms under such conditions would retreat to lower levels, out of harm's way, until the soluble salts became 'fixed' by the soil or converted to other materials by the soil microbes. This would take place within two to three weeks. There is no doubt that any strong solution would be distasteful to worms whether organic or inorganic in origin. So that, providing frequent top dressings of soluble salts are not given, the worms will not suffer seriously. Even so, there is no denying that worms do disappear where much artificial manure is used. The most obvious reason is that organic matter will be low under such conditions, and this, directly or indirectly, is the food of worms.

It has been shown before that the addition of nitrogenous matter, organic or inorganic, stimulates the soil organisms to destroy soil humus. Thus, if the crops leave little fibrous matter in the soil, and most farm and market-garden crops do not, a steady decline in the humus content will occur. It is this humus or peaty material which enmeshes the soil particles in a gelatinous sponge, and prevents it slipping into a semi-liquid condition under wet conditions. The gradual disintegration of this colloidal humus leaves the soil particles without support, giving a muddy mess in wet weather, and brick-like hardness when dry. Such soil is, in fact, little different from subsoil. This is probably the most obvious evil of the use of artificials and of much more consequence than the presence of sulphates. Our authorities are well aware of it. Hence the present drive for more organic matter in the soil.

Sir James Scott Watson, Chief Scientific and Agricultural Adviser to the Ministry of Agriculture, spoke of his concern at the declining humus content of soils at Seale Hayne College recently. Dr. A. M. Smith of Edinburgh, a soil scientist, also spoke of the excessive use of fertilizers to replace good husbandry.[1]

Dr. I. F. Storey, Plant Pathologist, Eastern Counties, speaking at a Horticultural Education Association conference in 1953, considered that the effects of poor soils and waterlogging caused much trouble with crops, and conditions suggesting disease could be due to no specific cause.

[1] *Farmer and Stockbreeder*, 7th April 1953.

The Problems of Present Methods of Agriculture

As long ago as 1941, Dr. Cleveringa, State Agricultural Consultant for Research for Holland on the Structure of the Soil, published a pamphlet drawing attention to the deterioration in soil structure occasioned by heavy dressings of artificials. He also advocated the application of farmyard manure to pastures in early August when the grass can make use of it, and deprecates winter application.

Coming nearer home, the potato land of south Lincolnshire, when first used for this crop some forty-five years ago, yielded 25 tons per acre, and very moderate quantities of artificial manures were used. Nowadays, with an application of nitrogen equal to ten times that taken out by the crop, their yield is only 12–15 tons.

These large crops, stimulated by the use of artificials, on good soil, were not maintained—neither was the organic matter. No amount of artificial fertilizer can compensate for this loss.

This collapse of structure, of course, leads to compacting of the soil and loss of that very vital soil function, aeration, without which plants will not grow. Hence the urge for deeper ploughing to create artificial aeration. Unfortunately, this artificial structure either dries into hard clods or slumps into a semi-liquid mass with the first heavy rain, and, as a result, it is difficult or impossible to get a tilth. Microbial and worm activity is lost and no plant foods are released by natural agencies. Such artificial manures as are added are rapidly available for a short time but soon become fixed or are attacked by the denitrifying bacteria (which can work without air), so that long before the crop matures the supply is exhausted, however much is put in. This, of course, represents an extreme case, but there are many soils approaching this condition, too many for the health of present-day agriculture. The process is so slow most farmers do not notice it, but if the farmer were to study the soil in the hedge-bottom and consider what he would save in cultivations alone if his fields were in the same condition, he would be eager to find any economic method to restore his soil.

If only for the foregoing reasons, it will be seen that our attempts to grow crops by a system of 'hydroponics in soil' are doomed to failure. By adding easily soluble plant foods to the soil we are not, in fact, helping the plant at all, we merely get the worst of both worlds. A fertilizer, in the full sense of the word, renders the soil more fertile. To call nitrogenous manures fertilizers is a misnomer, they decrease long-term fertility—they are, in fact, merely a stimulant and need to be recognized as such.

It will be noted (see graph) that after a rise in cropping for the first

THE INFLUENCE OF MANURES UPON SOIL FERTILITY

(From Russell after Waksman)[1]

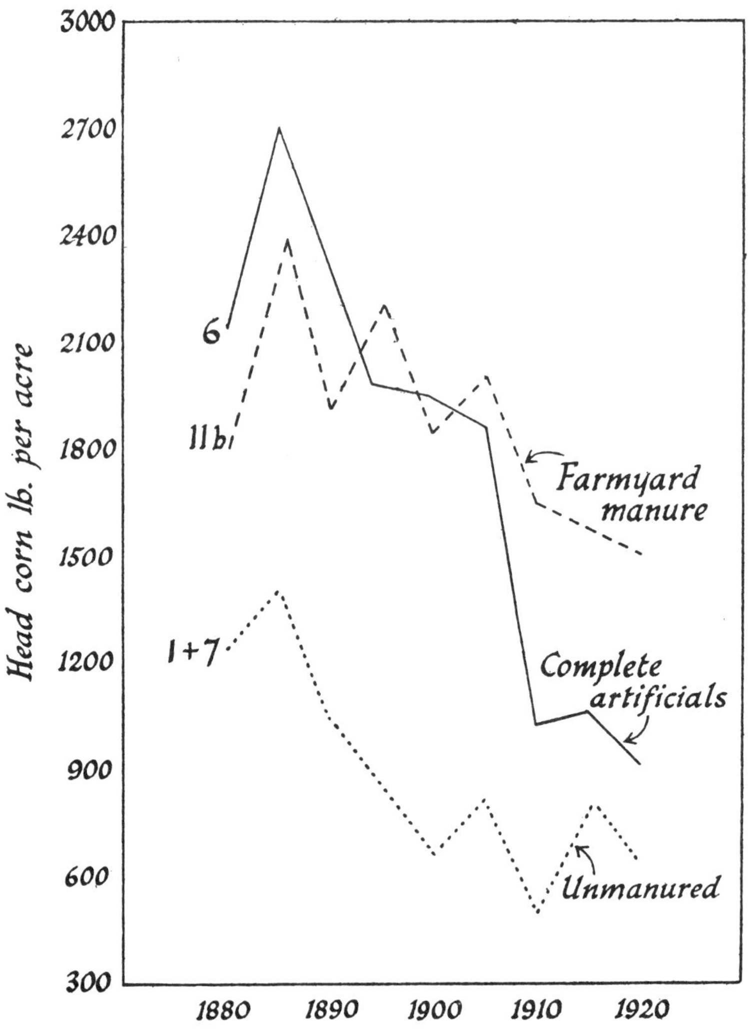

[1] *Soil Microbiology*, 1950, p, 311.

Plate 11

EFFECT OF GRASS ROOTS ON SOIL STRUCTURE

*

Soil samples from adjacent plots:
Left, arable for many years, only crop residues returned to soil.
Right, the same, but the last three years under Chewing's Fescue grass.
(*a*) As cut from the soil.
(*b*) Exposed to approximately 6 inches of rain showing the collapse of soil
where few roots are present. The turf sample shows little change.

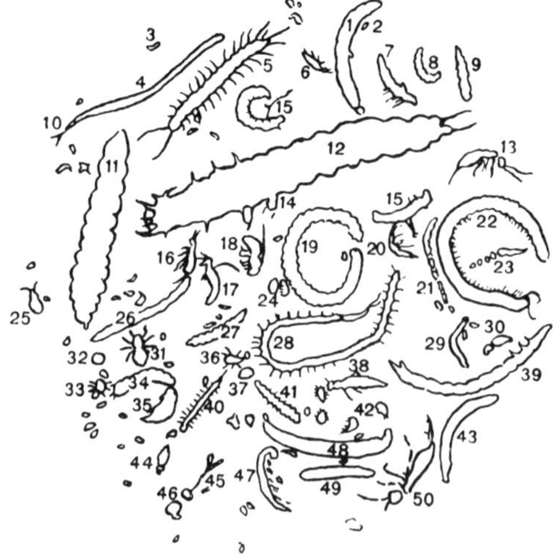

Plate 12

SOIL FAUNA

*

CHILOPODA: *Centipedes*
      4, 5, 28, 40.
DIPLOPODA: *Millipedes*
      19, 22, 41.
ARACHNIDA
(*a*) Araneae: Spiders; 31, 33.
(*b*) Acarina: Mites; 2, 14, 36.
Gamasidae, 24, 32, 37, 45:
Oribatidae; 42: Tyroglyphidae

INSECTA: *Insects*
(*a*) Collembola: Springtails; 3, 6, 10, 13, 16, 20, 50.
(*b*) Lepidoptera: Moths; 8 (larva).
(*c*) Coleoptera: Beetles; 7, 9, 17, 30, 35, 38, 44: Staphylinoidea (adults); 15, 23, 27, 29, 34, 41: Staphylinoidea (larvae); 25: Carabidae (adult, small species); 12: Carabidae (larva, large species); 21, 26, 39, 45: Elateridae (larvae) Wireworms.
(*d*) Diptera: Flies; 1, 43, 48, 49: Bibionidae (larvae); 11: Cyclorrapha (larvae); 18: Anthomyidae (adult).   (*By permission of Rothamsted Experimental Station*)

few years there is a rapid decline—particularly where artificial manures are used; where, in fact, the final level is nearly as low as where no manures at all are used, and well below the crop of the original soil alone. The use of farmyard manure maintains the fertility fairly well, but even here a gradual decline has set in once the readily utilizable organic matter has gone. Cropping under organic methods, maintaining a high level of organic matter, would be at a level between that of the unmanured and farmyard manured in the early days of the graph.

The total quantity of organic matter in an average arable soil is 20–30 tons dry weight (compare this with the 60–80 tons of old pasture or woodland). This will hold up to ten times its weight in moisture and contain 1–1½ tons of nitrogen; so it is possible to farm without adding organic matter, other than plant roots, etc., for some years. The damage, however, is progressive and cumulative. This quantity of organic matter will hold a great amount of water in itself and, by preserving a good soil structure, permit a great deal more to be held in the fine soil cracks.

It is for this reason of nitrogenous manure stimulating microbial activity that adding dung to the soil is a relatively poor method of increasing its organic content. Dung contains a great deal of nitrogen compared with its carbohydrate content. It seems that, to give a microbe a balanced diet it must contain between 1·2–1·8 parts of nitrogen to 100 of carbohydrate (pages 74 and 89). If there is less nitrogen, there will not be enough to spare for the crop—the bacteria, having the first call on it, can increase much faster than the plant. If there is too much nitrogenous matter it will be wasted either by washing out of the soil or it will be destroyed by denitrifying organisms.

Farmyard manure, besides containing 2 per cent nitrogen (dry weight), consists largely of humified materials and relatively little carbohydrate. The organisms feeding on it continue to break it down until only the very resistant portions are left.

As shown before, to maintain arable land in good heart it requires at least 15 tons per acre per annum—simply to provide sufficient of this resistant peaty substance to keep the soil in good condition. It is, in fact, a relatively expensive method of spreading peat and plant nutrients on the soil. Most of the nitrogen is wasted. Now the futility of adding further nitrogen to land already dunged is obvious. It is a common practice in market gardens. I did it myself for three years, and was disturbed to find that after adding dung at the rate of 60 tons per acre, with 10 cwt. of artificials, annually, the ground was still as 'dead' as at the beginning. The position was, of course, aggravated by heavy rainfall, but it is not a sandy soil. The facts were put to a provincial soil chemist of the N.A.A.S. He had no explanation.

It seems, then, that the economic way to use dung is to spread it on grassland, or a mature woody green crop, or straw, and work it into the surface, when surplus nitrogen will be quickly built into the grass or into microbial protoplasm. To apply dung in small quantities to grassland grazed bare, and then turn it in, as is customary with most organic farmers, is a good method, but the more dung is applied the greater the amount of dead-grass fibre required to absorb the nitrogen in it.

This method of fertilization is by far the most economic for the farmer, as the labour of mixing compost is avoided. Though, of course, it would certainly be easier to make compost, and spread it, than spread straw and dung separately where these materials are available. As the bacterial action takes place in the soil there will be much greater microbial action on the soil particles, with a greater release of plant nutrients and less risk of nitrogen losses than when compost is applied to the soil. Figures given by Rothamsted show crop increases from 15 to 25 per cent where the straw and nitrogen were ploughed in over that composted. No doubt the nitrogen in this instance was synthetic and the results need interpreting with caution.

Green manuring is popularly believed to be a good method of building up soil fertility. It can be, but, as practised to-day, it is anything but. The ploughing in of a soft, young, green crop adds a large amount of nitrogen to the soil relative to the amount of carbohydrate or plant fibre. Some farmers even add extra nitrogen to rot it down. As shown already, this results in a net decrease in the amount of organic matter in the soil. To be effective, the green crop must be allowed to mature. There is, of course, a risk of nitrogen starvation of the following crop if ploughed in, but this can be overcome by surface cultivation, as explained elsewhere. The effect of the popular ryegrass is merely to remedy in part the detrimental effects of artificial manures on soil structure, by its remarkably tough root system which prevents the soil from collapsing during the cultivation for the following crop. It does not add anything in the way of plant nutrients nor, if ploughed in young, as is usual, does it maintain the level of soil organic matter. Rothamsted have done work on ryegrass in comparison with clover and fallowed land. Taking the fallow as standard, clover gave an increase in a crop of kale of 22 per cent, while ryegrass gave a decrease of 27 per cent. Mustard and tares in previous experiments made little difference to the succeeding crop.

The Problems of Present Methods of Agriculture

In view of the leguminous nodule bacteria's need for carbohydrate to fix nitrogen, the practice of close and frequent mowing of grassed orchards becomes open to question. It is said that when the clover in the sward is cut the carbohydrate supply to the root is cut off and the nodules decay, releasing nitrogen for the grass and trees. Be that as it may, by frequent checking of the clover the total amount of carbohydrate manufactured is very much reduced and may only be sufficient to repair the clover plant itself. The amount obtained by the nodules will, in any case, be much reduced and so consequently will the fixation of nitrogen. Add to this the relatively shaded condition of the clover under orchards, and it will be seen that the amount of nitrogen fixed may be very small—perhaps none at all.

While exact conditions of management must be found by experiment, it seems likely that mowing should only take place when the grass is getting ahead of the clover—say, once a month in summer. As a supporting example, the roadways on my market garden, sown permanently to Kent wild white clover, are cut about three times during the summer, with the result that the grass has been largely exterminated and the clover—which is normally considered to grow only 3 in. high—grows 6 in. to a foot in height. Incidentally, the amount of organic matter removed at each cutting is considerable and is a valuable nitrogenous material to be used elsewhere (see also page 149).

So also does the practice of applying nitrogenous manures to grass-containing clovers. One of the clover's adaptations to survival against competition from grasses, which are powerful extractors of soil nitrogen, is this power of supplying its own nitrogen. If a nitrogenous manure is used on the sward, this advantage is nullified and the clover will be checked or even exterminated, apart from the detrimental effect on soil calcium of some forms of nitrogen, an element much needed by clovers. In addition, the clover is the most nutritious part of the sward crop—so that the herbage value is lowered.

Probably the most firmly entrenched practice in agriculture is ploughing. Some of its problems have already been referred to, but there are others. Comparing the plant-root's conditions under normal cultivations with that of natural conditions, there is, instead of the compact mass of granules perforated by air channels of the natural soil, a heterogeneous collection of lumps, either touching at the corners, with air spaces between or well rolled down with air spaces obliterated. In the first instance no water can rise, in the second no air can descend, and very little upward movement of water takes place owing to the fine cracks being broken up. Even with favourable weather, giving a reasonably well-cultivated soil, the upward movement of water will not occur as the finer capillary cracks are destroyed. There is, instead, a series of holes and solids through which water can only pass downwards under its own weight. There can be no capillary movement in any direction. A dry soil in spring is inevitable.

A layer of dead plant-remains lies under this, unable to decay if cold or wet, and then, with the advent of warmer weather, rotting rapidly if air can get to it, or putrefying if it cannot. This results in a sudden, unsustained release of plant food or a release of toxic substances. In any case, the food is largely out of reach of the fibrous feeding roots, which develop best under the well-aerated conditions near the surface.

Surface temperatures are erratic and the subsoil remains cold in spring, while losses of organic matter are heavy.

The most damaging effect of ploughing is the plough pan.

It hinders or prevents the drainage of water. It stops the roots reaching the subsoil for moisture and minerals. It effectively stops air getting to the subsoil, and exterminates the deeper-burrowing types of earthworm. The cause is not primarily the plough but the weight of the tractor wheel running in the furrow year after year at the same depth.

The ill effects of panning are becoming more widely recognized by our advisory services, especially at higher levels, but the damage is not yet well appreciated by farmers generally. Their reaction, generally, is to plough a little deeper, necessitating heavier and more powerful tractors, which merely perpetuate the trouble. Most farmers, too, now take it for granted that more rain is required after a week of dry weather, whatever their soil, whereas the digging of a few trial holes would show them abundant moisture below plough-level. Provided the subsoil is not of a gravelly nature, it will contain the equivalent of $1\frac{1}{2}$ to 2 in. of rainfall for every foot of depth, which would last the crop a fortnight of summer weather. Even a prolonged drought of several weeks should not seriously inconvenience a crop on the great majority of our farmland soils, with depths of 3 ft. or more. The strength of growth of vegetation in hedge-bottoms and waste corners, in spite of intense competition from hedgerow trees, provides a useful indication of potentialities. My own soil was well panned when taken over, and it was puzzling at first to find that plants ceased to grow after a week of dry weather, in spite of a moist subsoil. It was noticed, too, that deep-rooted weeds and plants—docks, rhubarb, carrots, parsley and fruit bushes—seemed to flourish comparatively well. After a few years of surface cultivation, and the encouragement

MARKET GARDEN CROPS FROM A SURFACE CULTI-VATED SOIL WELL SUPPLIED WITH ORGANIC MATTER

(*a*) Tomatoes under cloches; 'Amateur' July 26th, after sweet clover. Most of first trusses already picked.

(*b*) Strawberries, second year 'Cambridge 582' 8 tons per acre. Mulched spring with compost at 30 tons per acre.

MARKET GARDEN CROPS FROM A SURFACE CULTI-
VATED SOIL WELL SUPPLIED WITH ORGANIC MATTER

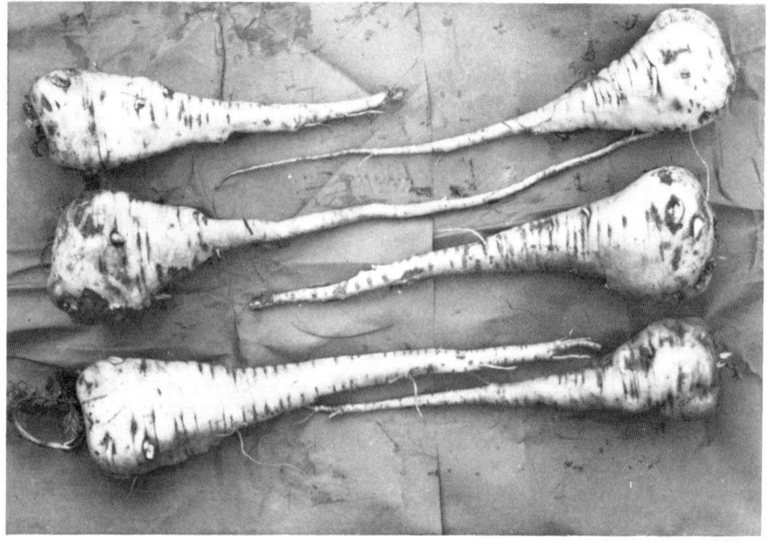

(*a*) Parsnips, averaging 1½ lb each. 30 tons of compost per acre.

(*b*) Coldhouse lettuce. 30 tons of compost per acre as a mulch.

of worms, and the use of deep-rooted plants, this pan became well perforated although a degree of hardness exists to this day.

It is now fairly well established that deep ploughing has no value over shallow ploughing. It may improve aeration of the surface layer on a wet soil by providing furrows to lead water away more easily and prevent the capillary movement from below. Such soils are unsuited to ploughing, in any case, and will suffer severe damage by panning with wheeled machinery.

In dry seasons it may reduce the yield. Rothamsted found that potatoes were reduced by almost 1 ton to the acre, presumably owing to excessive drying out of the soil (1951 and 1952). The work being done there on deep ploughing (12–14 in.) versus shallow ploughing (6–8 in.) for seven years has shown no difference as far as yields are concerned. It will be noted that the 'shallow' ploughing is not the shallow ploughing recommended by the organic advocates of 3–4 in. In a further experiment a comparison between plough and harrow, cultivator and harrow, and rotary cultivators, at two depths, 3–4 and 7–8 in., showed no consistent difference in yield between depths. Ploughing did show some improvement over the other two forms of cultivation. Probably, they suggest, because of better weed control in certain crops (but see Chapter VI). Further experiments done in 1944–5 included subsoiling and very deep ploughing for potatoes and still no differences were recorded, except one very wet field was improved by deep ploughing. It should be remembered, too, that their 'fertility' does not depend on organic matter.

It has been shown conclusively that hoeing does not have the effect on soils often attributed to it. The main purpose is to control weeds, and even then, in the case of deep-rooted crops like sugar beet, there is little advantage to be gained once the crop has covered the ground, provided there is sufficient available plant food.

Take the following experiments:

EFFECT OF HOEING ON BEET[1]

Tons per acre

	No Fertilizer	4 cwt. Sulphate of Ammonia
Intensive Hoeing	14·7	18·8
Normal Hoeing	12·4	19·5
Benefit of Extra Hoeing	2·3	—0·7

[1] Figures derived from E. W. Russell, Rothamsted Expt. Sta. Rep., 1949, 133.

The Problems of Present Methods of Agriculture

That it was only in the early stages that weed competition was harmful was shown by the following experiment.

EFFECT OF WEEDING ON BEET[1]

Tons per acre

Sulphate of Ammonia Given	Minimum Weeding	Intensive Weeding till 3 weeks after singling	Intensive Weeding throughout season
None	6·3	8·7	9·6
2 cwt. per acre	8·9	12·2	11·8
4 cwt. per acre	12·9	13·4	14·0

There is not much difference between the last two columns or where there was plenty of available nitrogen.

EFFECT OF HOEING ON POTATOES[2]

Total yield tons per acre (mean of 1937–8–9)

Hand Weeded	Frequent Hoeing	Weedy
10·9	10·6	9·0

Further experiments showed that five hoeings between potatoes were slightly more harmful than two hoeings. The damage to the roots and drying the soil were more harmful than the weeds.

Similar effects were shown in the case of a sensitive market-garden crop, lettuce.[3] The crop increased in weight, numbers, and individual size if kept weeded, but it made little difference whether it was hand weeded or hoed. A difficulty that may arise, however, is that in warm, wet weather a surface crust may form, and, especially when the soil has been heavily manured, much carbon dioxide will be formed and the plant roots become suffocated. So that a weed-free field, unhoed, may not crop so well as a similar weedy field hoed or hand weeded on account of the opening up of the surface by the two operations. Though it is not very likely to happen where there is a good worm population.

Soil is analysed by extracting it with a weak organic acid, usually citric. The quantity of potash, phosphate and lime leached out is claimed to be that available to the plant. No one yet has claimed that

[1] E. W. Russell, Rothamsted Expt. Sta. Rep., 1949, 134.
[2] Ibid., 132.
[3] Ibid., 136.

a plant, in fact, uses 2 per cent citric acid to extract its minerals, so that, although there is undoubtedly a concentration of carbonic acid around the roots, the method is quite arbitrary, and the results need interpreting with caution.[1] Soil analyses are an indication of potentialities rather than actualities.

I have found myself that a soil may have a low analysis of phosphate and potash and yet give a good crop. Five-year-old blackcurrants, for instance, gave a $3\frac{1}{2}$ tons to the acre crop with a soil analysis very low in phosphate and low in potash, with a lime requirement of 20 cwt. of limestone per acre. Five years previously it was very low in both potash and phosphate. All nitrogen, potash and phosphate, therefore, either came from or was released by a 3-in. mulch of straw, applied annually for the last three years.

A further interesting example occurred during an experiment over some years concerning the availability of potash. A plot of ground known to be very low in potash was sown to mustard early in 1950 and the crop allowed to mature. The litter formed was lightly cultivated in the following spring with 2 oz. of hoof and horn per sq. yd. Lettuce was sown for August cutting, and these were intersown with carrots at the end of June. Both did well, the carrots turning out at 12 tons per acre. The lettuce seed had been mixed with sweet clover seed and the latter established itself after the lettuce was cut. An analysis by the National Agricultural Advisory Service the following spring (1952) showed a potash level of 0·001 per cent, which is very low indeed. The clover was allowed to mature and die back as a 'sheet compost', and a further analysis in spring showed available potash as 0·006 per cent—an appreciable rise but still very low. This litter was again lightly cultivated in and planted to early potatoes (Home Guard), a crop needing much potash.

In order to test for the availability of potash as a limiting factor in cropping, the plot was divided into 12-sq.-yd. sections, some unmanured, one given sulphate of potash and another ammonium phosphate, both at 2 oz. per sq. yd.

Most plots, including that given potash, gave just over 60 lb. of potatoes, i.e. 11 tons per acre. That given ammonium phosphate gave over 90 lb., i.e. $16\frac{1}{2}$ tons per acre, probably due to the extra nitrogen, as the phosphate analysis seemed satisfactory.

In spite of the very low level of potash, as shown by analysis, there was, apparently, ample for the crop. A further analysis at the beginning of July—when the crop had matured—showed a potash

<hr>

[1] Rothamsted Expt. Sta. Rep., 1939–45, 51.

level of 0·011 per cent in an unused plot and 0·009 per cent in the unmanured potatoes.

While figures from single plots cannot be taken too literally, a rise in potash from 0·001 to 0·011 per cent represents an addition of about 4 cwt. of sulphate of potash, all of which remained available.

On the other hand, another area on which much dung and artificials had been expended, gave an analysis very high in potash and phosphate, yet failed to grow cabbage plants, as most of the nitrogen and organic matter had been lost.

Another interesting case of a steady rise in availability, as measured by the soil analyst, is given by Mr. D. Yellowlees.[1] No minerals were added except rock phosphate and basic slag in a few fields, and yet the whole farm showed a considerable rise in available phosphate and potash simply through using organic methods. Mr. Friend Sykes quotes similar experiences,[2] while the results of the Haughley experiments have already been quoted.

Soil analysis from my own experience gives a pretty fair estimate of the artificial manures added in previous years, but a low reading does not necessarily condemn the crop, provided adequate organic matter is present.

A plant's nutrient requirements are often diagnosed by analysis of the plant itself. The plant, though selective to a large extent, tends to take up whatever is dissolved in the soil solution though not necessarily utilizing these substances in its tissues. Therefore, the mineral content of a plant may vary according to the type of soil. Experiments by the Florida Experimental Station[3] have shown that plant composition varied little according to the manures added, or results of soil analysis from place to place on a particular soil type, but varied considerably according to soil types, i.e. acidity and organic content. It seems, then, that the composition of the soil solution depends much more on the type and condition of the soil than the nutrients applied. So that to apply fertilizers according to the needs of the plant, as shown by analysis, may not produce the desired effect.

In the case of glasshouse soils, however, very large quantities of fertilizers are added, sometimes more than the soil can render insoluble, so that normal nutritional activity is completely upset by

---

[1] *Journal Soil Ass.*, Vol. 5, 3.
[2] *Humus and the Farmer.*
[3] 'Factors Influencing the Mineral Composition of Vegetables', Florida Expt. Sta. *Bulletins* 438 and 488.

the salty soil water. Thus, the analysis of glasshouse tomatoes, etc., needs interpreting with caution also.

On account of these effects of the soil on the plant, analysis of the fruit or the seed of the plant is the safest guide to their needs. This varies little from plant to plant, wherever it is grown; it is, in fact, typical of the plant.[1,2] On the other hand, the root analyses is more akin to that of the soil, with the leaves and stems in an intermediate position.

From the organic farmer's point of view an analysis of real value would be that of the total nutrients present in the soil and subsoil. He would then have an idea how long his store of phosphates and potash would last him at his present rate of consumption, or, alternatively, the rate of consumption equivalent to the rate of breakdown of the underlying rock, that would permit him to go on farming indefinitely.

Of even more value would be an analysis of the total organic matter, total nitrogen and its C.-N. ratio. This would give a clear picture of the quantity of biological activity to be expected.

The position was well summed up by Mr. A. H. Hiller, J.P., addressing a convention of the Association of British Organic Fertilizer Manufacturers.[3] He pointed out that the trials of organics versus inorganics at research stations showed no advantage from the use of organics. Yet most market gardeners used large quantities of the expensive organic manures and went on to explain this by saying: 'The soil is a living organism. It can be analysed chemically just as you and I, after we are dead, can be analysed chemically, and it is a rather humiliating thought that you and I, like the turnip, are largely composed of water. Just as the chemical analysis of the human being, or the turnip, gives little indication of the living organism, so the chemical analysis of the soil can be equally misleading.

'Soil, like a human being, can be healthy or ill. When the soil is well it is either neutral or slightly alkaline, has abundant humus and differs slightly chemically from a sick soil. The difference is not great chemically but is very great physically.'

There is one aspect of organic methods which is of great theoretical and practical interest. Many of our most successful organic farmers —Newman Turner and Friend Sykes, for example are—getting very

---

[1] G. I. Nightingale, *Bot, Gaz.*, 1942, 103, 409; 1943, 104, 191; *Soil Sci.*, 1943, 55, 73.

[2] D. I. Arnon and D. R. Hoagland, *Bot. Gaz.*, 1943, 104, 576.

[3] *Fertiliser & Feeding Stuffs Journal*, 11th May 1948.

good yields from land that previously had large quantities of artificial manures applied to them. To what extent are they in fact releasing plant nutrients previously fixed? These residual effects, in fact, may go on for a long time, as the following figures from Rothamsted show.[1]

Manuring Prior to 1901 None Since	*Barley*		*P2O5% in Crop*	*Soil Analysis*	
	Grain cwt. acre	*Straw cwt. acre*	*cwt. acre*	*N. per cent*	*Avail- Phos- phate*
None	11·0 (8·8)	15·2 (10·4)	·104 (·047)	·110	1
Nitrogen only	11·4 (10·8)	13·6 (13·9)	·098 (·056)	·112	1
Superphosphate	20·9 (25·3)	22·4 (24·8)	·206 (·155)	·112	6
FYM	22·2 (26·4)	24·0 (24·8)	·243 (·165)	·138	6

Figures for 1949 and 1950.   (1949 in brackets).

These effects were even more accentuated during the dry spring of 1949.

It will be noticed that even after fifty years of continuous barley, with no manuring, the crops on those plots which received superphosphate or dung for many years previously gave about double the yields that the unmanured or nitrogen-only plots gave. Probably due, it seems, to the gradual release of phosphates, if the figures of easily available phosphates are any guide. It will be noted that there is six times as much easily available phosphate in the phosphate-and-farm-yard-manure plots as in the umanured and nitrogen-only plots.

Nitrogen, however, shows a different story. It will be noted that there has been little if any residual nitrogen where artificial nitrogen was used, which is in line with the facts already given; but in the case of farmyard manure there is a significant increase. This will be due to the 'hard core' of combined nitrogen in the resistant humus or peaty material left after the decomposition of dung.

This table raises an interesting point in view of the previous statement that only 20–30 per cent of the phosphates are ever recovered. Rothamsted themselves raise the point and conclude that perhaps the rate of recovery is higher than has been believed hitherto. The differences between the plots had been visible for many years since the cessation of manuring fifty years ago, but it is only of recent years that the differences have become so marked. It seems likely that this

[1] Rothamsted Rep., 1950, 43; and 1951, 45.

accentuation is not unconnected with a recent change of treatment. The land had been given liberal dressings of inorganic nitrogen since 1940. This would speed up the destruction of organic matter. Furthermore, it is the more readily utilized forms of carbohydrate that will be decreased most markedly—food for organisms that could have released the pound or two of phosphate per acre required on those plots which have never received phosphatic manures. It may well be that the compounds formed in the soil by adding soluble superphosphate over many years are rather more easily available to the plant, or it may be simply due to their greater quantity and therefore greater opportunity for the plant to extract them. A dry spring would decrease bacterial activity and accentuate these differences between plots.

Furthermore, the intensified breakdown of organic matter by the addition of nitrogen will release appreciable quantities of phosphate, as organic matter contains 1–2 per cent in combination. The small reserves in the soil organic matter will be running out by now.

It is of interest, too, that the percentage of phosphorus in the crop should be two or three times as much in a soil well supplied with minerals. A plant, it seems, will not normally take up any additional nutrient if well supplied—but if short it can make do on much less. In 1950, for instance, the total crop of about 45 cwt., including straw, on the farmyard-manure plot, removed 13 lb. of phosphate, while that which had received no manure only removed about 3 lb., for a total crop of about 25 cwt. Weather, too, has a surprising influence on this soil (which is low in organic matter). In a dry season a crop of 50 cwt. removed only 9 lb. of phosphate from the farmyard-manure plot and only 1 lb. in a crop of 18 cwt. from the no-manure plot. Taking the two extremes, the crops have a difference of five times in their phosphate content, due entirely to soil and weather conditions. With the higher water and micro-organism content of a soil well supplied with organic matter such extremes of mineral content would be unlikely, and it is more likely to be at the higher level.

Continuous cereals, it seems, are a possibility on artificial manures only (as are other grasses) for several reasons. For one thing, about 10 cwt. of organic matter (dry weight) are added annually in weeds, stubble and roots, and the more fertilizer used the greater the amount of stubble and roots, within reason. They are also powerful extractors of nutrients, especially nitrogen, largely, no doubt, on account of their very fibrous root system. This root system also maintains a good soil structure. They have, moreover, a large store of nutriment in the seed which enables them to develop quickly in spring and so take advantage of the fertilizers before they are lost in various ways. The fresh seed used annually in such cases contains a reserve of trace elements which may be needed by the crops sufficient to alleviate any shortage that may exist.

It may not be possible on all soils but, providing the drainage and aeration is adequate and the soil retentive of moisture and well supplied with calcium (as in Rothamsted soil), there does not seem to be any limit. The 35 cwt. of wheat annually for the past 100 years from Broadbalk Field, Rothamsted, seem to prove this. Yields, however, are apt to fluctuate violently with the season. As regards the quality, especially feeding value, of the wheat in this instance, we have no evidence. It would be interesting to compare it with, say, Haughley wheat. As regards its disease resistance, too, we have no direct evidence, but organic farmer's corn is not so readily lodged even with more unfavourable climate and soil.

Sterilization of the soil is an effective, if temporary and expensive, way of restoring the capacity of our glasshouse soils to grow crops and incidentally kill off many diseases. Sterilization is not merely a matter of killing off pathogenic organisms. Fungicides, while effecting some control over the pathogens, do not release large quantities of nitrogen as do sterilizing agents. While the exact mechanism of this effect is not yet determined it is not difficult to propose one.

When steam or other disinfectant is used on the soil all actively living organisms perish. All that survive are the resting stages of some bacteria, most likely the more primitive kinds that withstood the rigorous conditions of the earth's early days. At any rate, the fungi[1] and higher micro-organisms do not seem to be able to survive, neither do nitrifying organisms.[2] This leaves the field free for bacteria who have an enormous amount of dead protein to feed on, and in such soils very little carbohydrate. This results in the release of much ammonia. It is possible, too, that the heavy ammonia production is due at first solely to enzymic action on the dead protein. This excess nitrogenous matter stimulates the bacteria to attack such carbohydrate as there is and also the humus in the soil—releasing further nitrogen. The result is an excessive supply of food for the plant at the expense of soil structure. It is interesting to note in this connection it is recommended that chopped straw or other easily available carbohydrate be mixed in the soil after steaming to absorb

---

[1] Rothamsted Expt. Sta. Rep., 1951, 41.
[2] Cheshunt Rep., 1952.

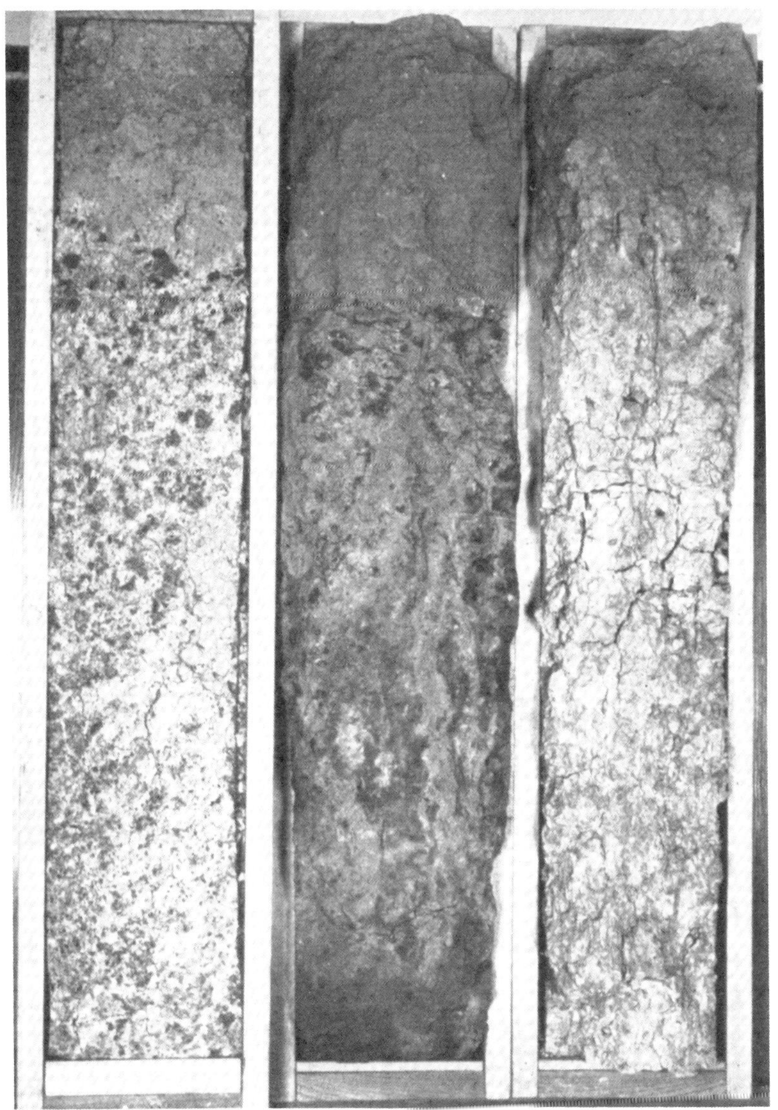

Plate 15

EFFECT OF CULTIVATION ON SOIL

*

The two samples on the left show the mottled markings indicative of water logging. That on the right though having a well-structured and drained subsoil, has, like the others, lost the structure of the topsoil. These would all be restored by laying down to grass and drainage where necessary.

(*By permission of Mr. B. S. Furneaux*)

EFFECTS OF ORGANIC MATTER AND CULTIVATIONS ON SOILS (II)

*

Photographs taken on the same soil type about 15 yards from one another.

(*a*) Ploughed, manured with fertilizers and dung. (Barley stubble). Showing the close-grained structure and consolidated layer, giving poor aeration.

(*b*) Rotary cultivated 2 inches deep, manured with compost or green manure crops. Showing course blocky structure and numerous worm-holes (After runner beans).

this surplus nitrogen. It may prove more effective than anticipated by forming appreciable amounts of humus if the above view is correct.

As evidence of the above proposition it may be mentioned that:

(a) There is a large increase in ammonia (five or six times),[1] though nitrates alter little, evidence of the decomposition of proteins in the soil population.

(b) A large increase in bacteria for some weeks, remaining above normal for six months or more.[2]

(c) Fungi are slow to return[3]—or may not return at all (particularly after steaming), showing that any readily available carbohydrate originally present has been consumed by the large bacterial population well supplied with nitrogen.

(d) A large increase in carbon-dioxide[4] production (about 50 per cent, further proof of increased destruction of organic matter.

(e) These effects decrease in the course of the year with the accumulation of readily available carbohydrate from dead crop roots and consequent reinfection with fungi. Where soil sterilization has been practised frequently it is found necessary to repeat the process annually and add quantities of nitrogenous manures as top dressings —the reserve of nitrifyable materials in the soil is so small that its normal rate of nitrogen release is negligible.

(f) Any disturbance of the soil leads to rapid resumption of nitrification of ammonia by reintroducing a more normal soil flora.

(g) Where organic matter has been added to the soil previously as composts or dung a large release of nitrogen occurs. No such effect occurs after adding inorganic nutrients and sterilizing.[5] Much of the nitrogen in compost or dung is in the bodies of micro-organisms which are destroyed by the sterilizing agent and the nitrogen released. For the same reason, on unsterilized soil, fertilizers will give more readily available nutrients than composts. An unfortunate result of sterilization is the almost inevitable loss of nitrogen[6] through heavy ammonia production and the increase in denitrifying organisms—the latter have been evident after twelve months.[7] This could be largely overcome by incorporating chopped straw.

[1] E. J. Russell and H. B. Hutchinson, *J. Agric. Sci.*, 1909, 3, 11; Rothamsted Expt. Sta. Rep., 1951, 42.

[2] Rothamsted Expt. Sta. Rep., 1951, 55.

[3] Rothamsted Expt. Sta. Rep., 1951, 55.

[4] Ibid., 41 and 42.

[5] Rothamsted Expt. Sta. Rep., 1950, 41.

[6] An instance of loss is given by Jens Roll-Hansen, Rep. State Expt. Sta., Norway, 1952, 10.

[7] Rothamsted Expt. Sta. Rep., 1952, 58.

The Problems of Present Methods of Agriculture

Sterilization by formalin is not quite so drastic as that by steam, in so far as it only produces about half the quantity of ammonia. Some organisms are fairly resistant to formalin, with the result that these increase rapidly, being freed from competition. The result of sterilization, though following the same general pattern, differs in degree according to the agent used.

It is often maintained that crop yields are lower than a hundred years ago when only organic manuring was practised. Certainly, when one considers the amount of breeding for bigger and better varieties that has gone on, one would expect them to be very much more at the present time than a hundred years ago. A few figures of Scottish crops from an old book by Sir John Sinclair, published in 1812, entitled *An Account of the System of Husbandry Adopted in the more informed Districts of Scotland*:

Oats now average 43 bushels, but under good conditions they obtained 58–72 bushels. A record crop was 87 bushels.

Barley rates now at 40 bushels per acre, the average then was 39–48 bushels.

Wheat was low, figures of 23–29 bushels are quoted against to-day's 39 bushels. However, it seems that the need was desperate and there were 270,000 more acres under cultivation than at present, so it seems that much of it was unsuitable land.

Potatoes were much the same as to-day, but turnip crops were phenomenal—to-day 18 tons per acre is average, while in those days 30–32 tons were usual—and up to 48 tons.

Considering that much of their seed must have been home-saved there certainly seems to have been a decline in yields in these cases.

Mr. Arthur Bryant, in his book, *The Age of Elegance* (pages 143–4), quotes some interesting figures from the period 1812–22. The Isle of Wight produced seven times more than the inhabitants could use, of a variety of crops. At Milton, in the Vale of Pewsey, an area of 3,500 acres, turned out annually 3,000 quarters of wheat, 6,000 of barley, the wool of 7,000 sheep, and eggs, milk and poultry. Cobbett reckoned that every land worker raised enough food for fifty to 100 people. This proportion has not been improved on to-day. This without the aid of artificial manures and mechanization.

Another calculation made recently on the records of American wheat crops[1] show that according to field trials of new varieties, those varieties should have given another 10 bushels per acre between 1910 and 1930, even by American standards of yield. Those average a good

[1] L. D. Bower, *Hawaiian Planter's Record*, 1949, 53, 1.

deal lower than ours. In actual fact, their increases only mounted to 3·2 bushels. How much did our yields increase between 1910 and 1930, if at all?

A difficulty which afflicts the farmer these days, especially the big user of artificials, is that of soil acidity. Usually this is synonymous with a shortage of lime. When an artificial manure in the form of a sulphate is added to the soil a chemical compound is formed with the lime in the soil—calcium sulphate. This is slightly soluble and washes out of the ground or accumulates in the subsoil as an inert material. This can still be used by the plant (exchangeable calcium) but has no influence on the degree of acidity of the soil. Only calcium carbonate (limestone or chalk) can do this and has to be replaced in quantities approximately equivalent to the sulphates used as fertilizer.

The soil may, however, be already low in lime—especially true of sandy soils and drained fenland. It may, in fact, be so acid as to inhibit plant growth. Worms are absent from such soils and plant residues accumulate as a mat on the surface, finally being converted to a peaty substance.

The application of composted wastes has been shown to enable plants to grow in such soils very well. It seems that the micro-organisms are able to extract calcium (lime) from the soil and hand it to the plant in some way, even though it may be present in very small amounts. It seems unlikely, however, that we could remove any great quantity of crops for long if the total amount of calcium in the soil is very small. This, of course, applies to any nutrient in short supply, and, as explained elsewhere, such soils would be better avoided for food production.

An illustration of clover plants grown in a very acid soil with and without compost is shown. The difference is startling. Liming, on the other hand, apart from remedying a real deficiency of lime, cannot confer on the soil any benefits that organic matter cannot provide. Even a deficiency of calcium can be remedied if the organic matter contains any quantity, and it usually does. Liming gives a ploughed-clay surface a more friable texture, but the way to avoid the necessity is obvious. It also speeds up the decomposition of organic matter in the soil, giving a greater release of plant foods—while at the same time appearing to maintain the soil texture. The organic matter is not inexhaustible, however; hence the expression that 'lime enricheth the father and impoverisheth the son'. This loss of organic matter, amongst other things, makes for trace-element difficulties quite apart from effects on soil structure. In excess, lime induces deficiencies of phosphates, potash, iron, manganese and magnesium. While liming may have advantages in special circumstances, it seems to be one of the many farming practices where its immediate effects have provided the evidence for wholesale acceptance. Its long-term effects, as in many other instances, have gone almost unnoticed.

An odd fact in this connection is that plants growing in culture solutions of very different levels of acidity quickly change the solutions to similar levels.[1] It may well be that a plant can alter its immediate soil environment also.

If, however, it is found necessary to crop a soil very deficient in calcium it is unwise to supply a large quantity of lime at a time, as this is apt to cause mineral imbalance. It is best applied as dressings of, say, a ton to the acre of a coarsely ground limestone which does not dissolve rapidly in the soil but provides plant roots with small particles of limestone from which they may obtain calcium as required. The use of burnt or hydrated lime, unless exposed to air for some time to convert it to the carbonate, is inadvisable on soil high in organic matter, especially if cultivated in, as it destroys proteins and nitrogenous matter generally, releasing ammonia which may be lost. Crops vary considerably in their tolerance of acid conditions—it may prove more profitable to use those crops adapted to such conditions than apply lime from time to time. Though under organic methods it will not be required, except perhaps at very long intervals, there being no acidifying fertilizers to counteract.

The effect of lime is nicely demonstrated by the following table.[2]

Treatment	Soil pH	Nitrogen Content per cent	Carbon Content per cent	Total Crops over 24 yrs.	CO_2 Production from 1 Kg. of Soil in 14 days
Minerals (P+K)+ Manure	5·5	0·146	1·7	69,300	999
Minerals+Manure+ Lime	6·7	0·143	1·7	59,754	1,100
Minerals+Sulphate of Ammonia	4·4	0·106	1·2	41,754	418
Minerals+Sulphate of Ammonia+Lime	6·1	0·095	1·1	61,906	523

[1] Rothamsted Expt. Sta. Rep., 1948, 41.
[2] S. A. Waksman, *Soil Microbiology*, 1952, p. 162.

The Problems of Present Methods of Agriculture

It will be noted that where lime has been used with manure, destruction of organic matter has been speeded up, as shown by the increased respiration of soil organisms, and, in spite of the more acid soil where no lime is used, the crop is about 16 per cent greater. Where sulphate of ammonia is used, however, the soil is very acid and a greater crop is obtained with the use of lime. The soil organic matter content of the last two treatments is very low. Soil nitrogen is also lower where lime has been used.

An approximate idea of the needs of crops in this respect is given by the following list, commencing with those preferring alkaline conditions and ending with those that will stand a considerable degree of acidity.

(Alkaline)

Lucerne	Barley	Red Clover	Rye
Sugar beet	Potatoes	Turnips	Swedes
Brassicas	Wheat	Oats	Timothy Grass

(Acid)

Allied to acidity is waterlogging of the soil through poor drainage, often aggravated by heavy rainfall especially on mountain land. This has the effect of dissolving out the calcium and some other minerals leaving a sour and impoverished soil. If this ground is drained, as it must be before any improvement can take place, lime and perhaps other minerals will be required to replace them. Only very small quantities of lime, as limestone may be added at a time to acid peat soils as the humus of which they contain a high proportion is rapidly nitrified and the nitrogen easily lost.[1] No doubt, in most cases, minerals would be gradually replaced from the underlying rock, but in the first instance some assistance is required by the plant. Deep-rooted plants are not very successful on such soils, generally, owing to inevitable periods of waterlogging during heavy rain.

In many parts of the world, especially where heavy or erratic rainfall occurs, erosion of the soil by water or wind is a serious matter. Most deserts, in fact, appear to have been the result directly or indirectly of faulty agricultural methods, and once started the progression may be cumulative.

As explained elsewhere, the finer soil particles are held together by the end products of decaying organic matter, while under natural conditions the soil itself is covered with trees or grass. Grass is a type of vegetation which can thrive when conditions are too severe at

[1] S. A. Waksman, *Soil Microbiology*, 1952, p. 189.

some part of the season for larger organisms to exist, either through extremes of drought, wet or exposure, and its existence under natural conditions should be a warning of a situation unsuited to arable conditions. Unfortunately, such vegetation, especially that growing under dry conditions, is particularly easy to clear and convert to arable. This has already occurred in recent times in central North America. Given suitable safeguards such land is capable of limited cropping and the safeguards adopted consist, in principle, of water conservation, the minimum of cultivations and leaving the organic matter on the surface to protect it. Such methods afford immediate protection and also check any further loss of organic matter. By such means much wind-eroded land in central North America has been brought back to cultivation.

Water erosion may be much more expensive to check and the consequences are more serious, taking a similar area into consideration. Such soils may be covered by grass or forest originally. When the protective vegetation is removed and the surface litter oxidized away, heavy rain beats the finer soil particles into the drainage cracks, and worm-holes where they exist, filling them up so that more or less rapid surface run off occurs according to the type of cropping, as shown in the following table.[1]

AVERAGE ANNUAL RUN OFF, 12-DEGREE SLOPE

Cropping	*Inches*	*As Percentage of Rainfall*	*Soil Loss tons per acre*
Continuous Maize	15	40·3	99·3
Crops in rotation:			
Maize	9	23·7	41·5
Wheat	9	24·8	11·4
First Year Ley	7	17·7	0·6
Second Year Ley	5	12·8	0·2
Permanent Pasture	1·6	4·3	0·02

It will be noticed how the bare-soil crops lead to poor absorption and rapid run off, resulting in heavy loss of soil. Protection by surface vegetation and as much as possible with shallow cultivations to keep the organic matter and better structured soil near the surface is the

[1] Zanesville, Ohio.

remedy. It may prove possible where rainfall is sufficient to undersow the maize crop with a legume.

Where run off has become serious, and formed gullies, then expensive reconstruction works are necessary, with the planting up of the eroded surfaces with quick-growing trees and plants, which are usually leguminous, these being able to support themselves in the absence of soil nitrogen. Provided there is at least a period during the year when the soil is permanently damp and animals and machines are kept off it, natural regeneration is a remarkably potent force.

Many people assert that artificial manures are harmful in themselves, i.e. synthetically fixed nitrogen compounds are not the same as naturally acquired ones. Therefore the proteins manufactured by the plant are not all they should be. J. E. R. McDonagh, in *The Universe Through Medicine*, expands this view and goes further to state his belief that this imperfect protein is the precursor of virus particles. Some of the arguments seem to be rather hypothetical though none the less interesting, and worthy of further study. The student of organic chemistry will appreciate the differences that can occur between compounds having the same chemical formula and very similar structure.

What may be a possible difference between synthetic and naturally produced nitrogenous compounds is in the relative proportions of the isotopes of the elements concerned. All elements, as far as we know, show this phenomenon of atoms having apparently the same chemical properties but slightly different atomic weight. It may be that their chemical properties, particularly when compounded into complex organic molecules, are not quite the same after all. Some plants have a decided predilection for these isotopes. Sudan grass, for instance, acquires appreciable quantities of 'heavy' carbon and nitrogen. Willows contain 'heavy' hydrogen. No work has been done, as far as I know, on the possibility, so the view must remain conjecture at present.

The view has been put forward that our plants have been bred so long under artificial conditions that by selection they are now adapted to them. It is very likely that by the survival of the fittest they are better adapted to such conditions than their ancestors. Whether they have come to prefer such conditions is another matter. It begs some very fundamental questions. It does not seem likely that even if the plant does not object to a change in the quality of its food, it could possibly benefit from the erratic meals which are inevitable under present methods.

131

VI

PRACTICAL ORGANIC FARMING

It is appropriate now to convert this mass of evidence and information to a few simple principles as a guide to its practical application.

Our problem is to utilize the natural elements of light, water, air and soil to our maximum advantage—which include our distant as well as immediate well-being.

The light and air are inexhaustible. Taking the total plant nutrients of the soil and subsoil and rocks beneath, they too are inexhaustible for practical purposes. Water also is inexhaustible but, unfortunately, local, temporary and widespread shortages do occur.

Our problem in more detail is to find means of extracting the soil minerals and to conserve soil moisture as much as possible. In point of fact the two go hand in hand.

In order to release these minerals the soil micro-organisms must be provided with energy food. With this they can manufacture all their other requirements—some one item, some another, the surpluses of one being absorbed by its neighbour. These manufactured products are absorbed by the growing plant on the death and disintegration of individual microbes.

The energy food of the microbes is manufactured by the green plant from the carbon dioxide of the atmosphere and water, with energy derived from the sun in the form of light. On the death of the plant all it contains is returned to the microbes.

There is, thus, a cycle driven by the sun, of material to and from the plant and soil micro-organisms.

It has been calculated that land plants over the world manufacture some 100,000,000,000 tons of sugar and other carbohydrates annually, giving some idea of the magnitude of the cycle. If we only harness a small percentage of this it will be all we require.

All other things being equal, the rate and magnitude of the cycle will depend ultimately on the amount of energy absorbed from the sun. And in order to absorb the maximum amount of energy from the sun we must have the ground covered with the largest possible amount of foliage to intercept the light at all times, using plants that have a high efficiency under the particular conditions of moisture and temperature obtaining. It is, in fact, this matter of quantity of energy absorbed that is the final limiting factor in agricultural production, and if agricultural crops could absorb it as efficiently as the natural vegetation there is little doubt our food supply would be increased materially. Consider for example the amount of carbohydrate manufactured annually in a mixed forest, from the soft celluloses of the herbs to the lignin of the trees.

Further, we must conserve this organic matter, the plant nutrients it contains, moisture and soil structure, by eliminating, as far as possible, disturbance of the soil. The practical measures to achieve these aims are, briefly, as follows.

Organic matter is built up by keeping something growing on the land at all times, by growing green crops after harvesting, by avoiding checks to growth, by autumn sowing where applicable, even a cover of weeds is doing something provided they are not allowed to get out of hand. Particularly is it necessary in autumn when mild and moist conditions allow great activity of soil microbes. At this time, if soil organic matter is too low for continued microbial activity, the heavy rain of winter will wash much of their valuable constituents out of the soil. Here again the forest or wild grassland gives us a clue—a carpet of dead leaves and plant stems in autumn—ideal food for the micro-organisms just when they most need it.

Secondly, cultivations must be shallow and infrequent to maintain as high a level of organic matter as possible.

With these two basic principles in mind, let us see how they work out on the farm and, putting first things first, how to convert a farm from present-day methods to organic methods.

It may be stated here, for those who fear a drop in yield, that this has not been the experience of most of those who have made the change. As far as my own experience goes, there was certainly no fall in yield but there was a decided fall in the incidence of pests and diseases, with a consequent improvement in quality, and saving in time and material combating them. There is, however, a risk of a fall in yield on land that has had its organic matter reduced to a very low level, more especially on wet, acid, or sandy soils where its lack is accentuated.

Labour changes depend on circumstances, if much stock is kept in yards, which is not recommended if avoidable, extra labour may be needed for composting. On a well-mixed farm little change need be anticipated, though cases are known where less was required even with much stock.

While the principle is not inflexible, it is usual to base the crops on an eight-year rotation—four years under a deep-rooted ley and four years under arable crops. The former to build up sufficient nitrogen and organic matter to support the latter.

A simple illustration of the effect of a ley on cropping, even a short one, in the rotation is shown by the following table derived from some Rothamsted experiments done without manures:

WHEAT CROP PER ACRE IN DIFFERENT ROTATIONS[1]

Rothamsted

Continuous Wheat	*Wheat Alternate with Fallow*	*Wheat in Four-Course Rotation (Swedes, Barley, Clover, Wheat)*
11·3 bushels	14·0 bushels	24·0 bushels

The type of ley used also has a bearing on the following crops. The following results show the effect of a one-year ley on wheat in cwt. per acre.

Rothamsted 1933[2]

Treatment of Ley	*Fallow*	*Clover*	*Clover and Ryegrass*	*Ryegrass*
One Cut	30·6	25·5	21·4	18·1
Two Cuts	—	25·1	18·9	13·1

Some old experiments at Rothamsted (when, no doubt, the organic content of the soil was higher than it is now) showed that an old lucerne or clover ley would benefit crops to a decreasing extent over five years.[3] It is estimated that the weight of roots in a clover ley

---

[1] J. W. and E. W. Russell, *Soil Conditions and Plant Growth*, 1950, p. 486.

[2] E. R. Orchard, unpublished Ph.D. thesis, London, after E. J. and E. W. Russell.

[3] H. Nicol, Emp. J. Expt. Agric., 1933, 1, 22.

exceeds the tops and contains quantities of gradually available nitrogen, potash, phosphate and calcium.[1] Grass roots, too, on account of their tough nature, help to prevent loss of soil structure for as much as three or four years.

The rotation starts, then, with the laying down of a four-year, deep-rooted ley, taking one-eighth of the farm each year. The ground is prepared for the ley by subsoiling, if possible, to break up the plough pan, and a shallow ploughing, incorporating, if possible, some organic matter as compost, dung or green manure. Or these materials may be disced in later. This may be done in spring or if the land is dirty with perennial weeds, a cleaning crop taken, or fallowed, and sown in August.

The seeds mixture contains a number of deep-rooted plants, to extract minerals, and leguminous plants to fix nitrogen, and the plants are chosen to give a palatable mixture to the stock for as long a period as possible.

A popular mixture, recommended by F. Sykes,[2] is the following:

    12 lb. Cocksfoot (Akaroa)
     2 lb. Italian Ryegrass
     3 lb. Timothy
     2 lb. S.100 Aberystwyth White Clover
     5 lb. Hants Late Flowering Red Clover
     4 lb. American Sweet Clover
     2 lb. Alsike Clover
     1 lb. Yarrow
     4 lb. Burnet
     3 lb. Chicory
    ──────
    38 lb. per acre
    ──────

This particular recommendation was for thin soil in an exposed position—so it is likely to succeed in most situations.

Another recipe recommended by F. N. Turner[3] is as follows:

GOOSEGREEN HERBAL LEY MIXTURE

    4 lb. Perennial Ryegrass (S.23)
    4 lb. Perennial Ryegrass (S.24)
    5 lb. Cocksfoot (S.143) ⎫
    5 lb. Cocksfoot (S.26) ⎬ on light or medium land
                            ⎭

---

[1] W. B. Goldschmidt and E. R. Orchard, S. Africa Dept. Agric. Pamph. 22, p. 140.
[2] *Humus and the Farmer.*
[3] *Fertility Farming.*

Plus/or 4 lb. Timothy (S.51) ⎫
         4 lb. Timothy (S.48) ⎬ on heavy land
                             ⎭
         1 lb. Rough Stalked Meadow Grass
         1 lb. Meadow Fescue
         3 lb. Late Flowering Red Clover (Montgomery or
               Aberystwyth)
         1 lb. White Clover (S.100)
         1 lb. Wild White Clover
         2 lb. Chicory
         4 lb. Burnet
         ½ lb. Yarrow
         2 lb. Sheep's Parsley
         1 lb. Alsike Clover
         2 lb. American Sweet Clover
         1 lb. Kidney Vetch
         2 lb. Lucerne
         1 lb. Plantain
         ½ lb. Fennel
         1 lb. Dandelion
    Plus 6 lb. Italian Ryegrass or 6 bushels of oats if sown direct, per
         acre.

This mixture, containing a wide variety of grasses and herbs, is considered to give a more balanced and palatable feed to stock than the restricted mixtures usually offered and, in addition, should one or two species prove unsuited to the conditions the loss will not prove so serious. It is also a good general-purpose ley (as is the previous one) and will provide good grazing or hay, as required. It will be noted, also, that most of the plants, particularly the legumes, are deep rooted and likely to stand drought well, at the same time opening up the soil and tapping a volume of minerals not commonly touched. There are, in addition, the similar but better-known Clifton Park mixtures.

This herbage may be lightly grazed in the autumn and winter, but as the purpose of the ley, particularly in the early stages of conversion, is to build up the organic matter of the soil and improve its texture and capacity for growth, this must be practised with caution, and in most cases will probably not be advisable. The following season it may be grazed; and the second, third and fourth season grazed or taken for hay and silage, or a combination, according to need.

By the end of the fourth summer there is a good build up of organic matter in the soil and it is ready for a series of arable crops, usually commenced with winter wheat. Wheat, of course, may come after another crop in a later part of the rotation, so it will be convenient to describe the sowing of this crop under both circumstances.

In the case of previous arable crops the ground will be covered with residues of the previous crop and weeds and their seeds. The first step, especially if weeds are numerous, is to cut them up and bury their seeds lightly to encourage germination. This should be done as soon as the crop is harvested by a shallow discing or rotary cultivation 1 to 2 in. deep. A carpet of weed seedlings will result which may be used as animal feed or left to provide green manure for the crop. Shortly before sowing, give a further one or two shallow cultivations according to the weediness of the soil. By this means weeds are destroyed before sowing instead of burying them with the plough for germination in years to come. A few more weeds may germinate after the corn is sown, but they will not be many. There will be a further batch germinating in the spring, but with the good soil conditions, provided by the cultivating in of the weeds the previous autumn, the corn will make rapid growth and exclude them. Rothamsted have shown that a good crop of wheat will smother weed in a fertile soil.[1]

It may be felt that the soil is never dry enough for rotary cultivation or discing so late in the year. This may be so in exceptional years and wet districts, but it will be found that soil under organic methods is remarkably well drained and never reaches the sodden stickiness of ploughed arable soil. In any case, the presence of fresh decaying organic matter from the weeds has a very beneficial effect on the soil tilth on account of the high fungal population which develops.

In the case of a ley being required for the crop, it is common practice amongst organic farmers to graze the turf bare, and rotary cultivate, disc or plough, perhaps with a light dressing of dung. Discing or rotavating in some wet districts may be impracticable on account of the impossibility of killing the grass, and a shallow ploughing of 4 in. just sufficient to turn the grass under may be a better preparation. Some farmers in such situations find it possible to break a ley with a disc by starting the cultivations in August and sowing the wheat in September—at 1–1½ bushels per acre. It seems, incidentally, that the common seeding rates of cereals are excessive. Rothamsted, growing winter wheat on land infected with eyespot, found that by sowing 1½ bushels to the acre, yield increased by 8 to 10 bushels per acre over a seeding rate of 3 bushels per acre.[2] The growth which ensues from such early sowing provides a useful late-autumn feed.

A fine seed-bed is not required at this time of year, there being

---

[1] Rothamsted Expt. Sta. Rep. 1939–45, 93.
[2] Rothamsted Expt. Sta. Rep., 1952, 91.

plenty of moisture in the soil in any case. Indeed, experiments on the yield and quality of malting barley seem to show that the condition of the seed-bed makes little difference for cereals.[1] This is one of the rare references to quality to be found in our orthodox experimental work.

After grazing off in spring it may then be harrowed and sown with a mixture of rye grass and trefoil, to maintain the earth cover the following autumn and winter and also to give a useful autumn feed after harvesting. Undersown trefoil, incidentally, is an excellent control for 'take all' where cereal crops are frequent. It also provides nitrogen for rotting straw after combine harvesting and is a useful check for annual weeds.

This is usually followed by a spring-sown crop: oats or barley, perhaps wheat again.

Again, the preparation will consist of two or three discings or rotavations, according to the weather, or, if conditions demand it, ploughing—shallow, as always—until the grass and weeds are destroyed. There will be no difficulty in providing a seed-bed in the soil by this time, providing elementary precautions are taken with regard to soil and weather. Further, providing the depth of cultivations is kept to a minimum—as may be achieved with a rotary cultivator, the seed may be drilled without subsequent rolling. Alternatively, it may be broadcast, harrowed and rolled.

After harvesting, the land should be sown to mustard or rape or even Italian clover if the crop can be harvested early enough. Other green crops may be used in mixture, ryegrass and tares, but anything likely to be difficult to kill off by shallow cultivations the following spring should be avoided, as it is at this stage in the rotation that small seeds are drilled—turnips, mangolds, sugar beet or kale. One need not, of course, be so particular if the land is required for potatoes.

All these crops will benefit from any available compost or dung—the former is much to be preferred, at 10–15 tons per acre, and on a mixed farm where this is an inevitable by-product, this is the best time in the rotation to apply it. For small seed crops it needs to be spread and cultivated in as early in the year as possible, say February or March, so as to allow ample time for the control of weed seeds. It should be pointed out that many species of weeds are unable to germinate until the warmer temperatures of May prevail—black nightshade and lamb's-tongue or fat hen being notable examples, besides

[1] Rothamsted Expt. Sta. Rep., 1949, 137.

that notorious weed of wet districts, persicaria or 'red shank'. If the land is known to be clean, sowing may take place towards the end of April, but good crops may still be gained from sowing in early June. Under the moisture-retentive conditions of organically cultivated land there is little risk of failure. With such late sowing, however, it is advisable to roll. There is no need to ridge the land for sowing these crops when there is a clean seed-bed.

Potatoes in one respect appear to present the organic farmer with a problem—how to get sufficient depth of tilth to draw up ridges. In fact, however, it is not necessary to ridge except in heavy or 'panned' soil. After the usual cultivation shallow drills 3 to 4 in. deep may be drawn out with any convenient implement, the potatoes planted and given their dressing of compost. The tops of the ridges can then be knocked down with the ordinary zigzag harrow used upside down. When the potato plants are 6 to 12 in. high they can be earthed up if weeds necessitate the operation.

Provided the soil is not hard below, or heavy, it does not make any difference to yield and little to greenness whether they are planted on ridges or on the flat, as shown by Rothamsted,[1] and frequent or deep tillage gives no benefit and may even be harmful.

A further point is the choice of potatoes for seed. It is hard to see why the average farmer chooses his small potatoes for seed. He does it with no other stock. Virus and diseased plants produce small potatoes, and one is thus perpetuating any disease with which it may be infected, and even the best stock almost invariably has a few individual infected tubers. My own Majestic seed has been saved for four successive years and they are just as heavy yielding, and certainly healthier, than the original certified Scotch seed. I did not find a single plant with virus infection or black leg in the past year, which is more than could be said of some certified Scotch Arran Pilot bought this season. This in spite of the fact that this is a potato-growing district. The reason is almost certainly the use of the largest potatoes, cut longitudinally to give each piece at least one eye.

It may be mentioned that the practice of chitting seed before planting does not seem to have the value popularly ascribed to it. While it has some advantage with very early crops and with late-planted crops, Rothamsted experiments show that under normal conditions there is only one extra hundredweight per acre to be gained from an April planting of chitted potatoes.

Beans are another crop which find a dressing of compost or dung

<hr>

[1] E. W. Russell, Rothamsted Expt. Sta. Rep., 1949, 137.

of particular benefit if available. Seed-bearing legumes have a particular need of organic matter in order to set seed. The clover-seed yields at Haughley, previously mentioned, are an example of this. The popularity of the bean, an important protein crop for those who feed their own stock, seems to have declined with the organic content of present-day farm soils. It is a crop which does well on organic farms. Rothamsted found that it did well on dung. This crop, too, may come at this stage of the rotation, but sowing takes place in the autumn after cultivating in the remains of the cereal. Compost is applied and disced in with the ensuing weeds. The seed is sown as soon as the land is clean, at 3–4 bushels per acre, and may be broadcast or drilled. This is a good crop for wireworm-infested soil.

The last year of the rotation can well be another cereal. The land is clean after the previous crop and it may be sown as soon as the previous crop is cleared. In the case of kale, this may not be until the following spring. Autumn-sown crops will follow the lines laid down previously for sowing after a ley, but it is unlikely that they will be sown early enough to provide so much spring feed. Whether autumn-or spring-sown, the crop will be undersown in spring with the four-year ley mixture.

This is an example of a rotation commonly practised by organic farmers. It is not the purpose of this book to give detailed instructions of farming practice and rotations; in any case, every farm is differently situated and each farmer must plan his own cropping to suit his needs. The purpose is simply to show how the two principles of keeping the ground under a green cover and shallow cultivations work out in practice. A farmer may wish to work in a permanent ley or field of lucerne or sainfoin, or grow market-garden crops or early potatoes—there is no more difficulty in fitting them to organic methods than to normal farming methods.

A few examples of rotations, however, may be useful.

For those whose cropping consists mainly of cereals, fertility can be largely maintained by sheet composting on combined straw or undersowing with clovers. Here are some examples:[1]

Barley	Barley	Barley	Oats	Oats
Wheat	Mangolds	Pasture	Barley	Wheat
Ley	Wheat	Hay	Beet	Beet
Wheat	Ley	Beet	Barley	Barley
Barley	Wheat	Barley	Hay	Wheat
Beet	Barley	Oats	Wheat	Ley

[1] I am indebted to Lt.-Col. Sir J. E. H. Neville, Bart., for these rotations.

On good land one-year leys are sufficient, but on extremes of sand or clay the full four years will be required, and even then, in some instances, it may be found that soil structure deteriorates too rapidly for four full years of arable.

Using the customary four-year ley:[1]

Four-year Ley	*Four-year Ley*
Beans and wheat	Kale, potatoes or roots
Oats	Oats
Kale, potatoes or roots	Oats and barley
Oats and barley undersown	Beans and wheat undersown

The second of these two rotations takes advantage of the fertility of the newly ploughed ley—provided wireworms are not a pest.

Another for poor ground:[2]

> Oats undersown—one-year ley
> Grazed or hay
> Oats undersown—long ley.

For better ground:[2]

> *Four-year ley*
> Kale
> Peas
> Wheat
> Oats
> Four-year ley.

On a medium loam:[3]

> *Four-year ley*
> Wheat
> Oats
> One-year ley
> Oats or barley undersown.

and finally from Mr. S. Mayall:

> Winter corn (wheat or dredge)
> Roots or silage followed by a folding crop
> Spring corn undersown
> Four-year ley.

[1] I am indebted to Lt.-Gen. H. R. S. Massey for these.
[2] To Messrs. C. and I. Howlett for these two.
[3] To Mr. R. Coward for this one.

One of the noteworthy features of the organic methods is the handling of weeds. They are always under control. By the simple expedient of not ploughing the soil, fresh seeds, often needing a spell of maturation underground where they may lie for twenty years or more, are not brought to the surface annually, shortly before the crop is sown. Opportunities for their destruction occur frequently, and at different seasons, making sure that those species which are late in germination are also destroyed. Gradually the top few inches of the soil becomes to a large extent exhausted of dormant seeds. Frequent opportunities for their germination and destruction are afforded by repeated cheap, shallow cultivations. It must be remembered that the cost of ploughing is equivalent to at least three discings —which in itself is sufficient to clean the land. After ploughing one has to scuffle, roll and harrow—perhaps repeatedly—and the result is still not so satisfactory.

The wild oat has become a serious pest of cereals in many parts of eastern and midland England of recent years. Shallow cultivations may well prove the best way to eliminate it. Indeed, it has proved so in Canada. This weed germinates when oxygen reaches the seed after rotting of the seed-coat, and if this occurs it will even push up from depths of 7–9 in. Rotting of the seed-coat will be slow at such depths and oxygen restricted in any case. If these seeds were not buried by ploughing annually, they would germinate much more readily and could be destroyed. It is found, too, that seeds only germinate in autumn, winter or spring—times of year when they can usually be dealt with. This combination of circumstance for the germination of seeds is common to many arable weeds: they cannot help but thrive under ploughing.

There is no need to fear a moderate amount of weeds in deep-rooted crops. The experiments at Rothamsted and Woburn have shown[1] that, providing the soil is well supplied with nitrogen and young seedlings were not smothered, the yield of sugar beet is not improved by weeding. Such weeds presumably comprise the normal weeds of arable land and not perennial ones. Potatoes and lettuce, however, were invariably reduced by even small amounts of weed— these crops deriving most of their moisture from the surface soil.

By the means outlined here one really is controlling the weeds—by using their peculiarities for their own downfall.

The disc, and to some extent the rotary cultivator, have come under criticism, accused of spreading perennial weeds. It is, of course, very

[1] Rothamsted Rep. 1939–45, 35.

satisfactory to some to see couch grass being dragged out in handfuls by a spiked harrow. The fact that this merely makes more room for the remainder to grow better is overlooked. One cannot kill perennial weeds by dragging them out; one merely thins them out, to their obvious advantage. Even the gardener working over a small patch of ground by hand never gets all the roots out in one operation. The multiplying powers of perennial weeds are enormous. Curtis, the botanist, once planted a few inches of creeping thistle root in his garden. In six months he had 4 lb. of root. The spread of weeds is a geometric progression, not an arithmetical one; that is, it does not merely add to itself, it multiplies itself. The only factor which prevents farmland becoming completely infested with such weeds is competition for light, moisture and food—usually with the crop. The answer, then, is to deprive it of light, moisture or food. The last two factors are virtually impossible to eliminate, but it is easy enough to deprive it of light by simply cutting its top off as soon as it pushes through the ground. There is probably no weed that can stand this sort of treatment for a full summer. It is certainly laborious, necessitating a fortnightly cultivation, for the first part of the season at least. The temptation to give up when the job is nearly finished is the undoing of most farmers. The final one or two cultivations are well worth while.

While the disc is a fairly good tool for this purpose, duck-footed or L-footed cultivator tines can do a good job properly set.

There is little doubt that the ideal implement for the job, and indeed most cultivation, is the rotary cultivator. The advent of this tool is one of the greatest advances of agriculture made, and will, I believe, oust the plough and traditional implements from most farms in the course of a decade or two, except in special instances. Never before has it been possible to produce a seed-bed from rough herbage with one implement, and so quickly. For the organic farmer it has the additional advantage that it can be used effectively at a shallower depth than any other tool. This is a particularly valuable quality for a seed-bed. My own experience is that even in the driest weather the soil does not dry out below the loose layer, and seeds germinate readily. F. Sykes, of Chantry, records a comparison between autumn-ploughed ley with the usual cultivations, and rotary cultivation 1 in. deep in one direction and 2 in. deep across this. The trial took place on the same field which was sown to oats in a dry April. The oats on the traditional seed-bed germinated in fourteen days, while on the rotary-cultivated seed-bed they came up in four.

It is as well to discuss here an experiment, already briefly mentioned, carried out at Rothamsted over a number of years, comparing cultivation by ploughing, tractor cultivator and rotary cultivator, at different depths, 7–8 and 3–4 in.[1] It was found that depth of cultivation made little difference to the crop but that, generally, ploughing showed a distinct superiority in regard to yield over the other two forms of cultivation. The deeper rotary and cultivator treatments gave a better yield than the shallow.

This result appears contrary at first sight to the methods outlined in this book. It seems, however, that the difference was due to severe weed competition for the crop in the early stages; in fact, it soon proved impossible to grow mangolds at all except by ploughing. The reason for this severe weed infestation is that the ploughing was followed by the usual cultivations—giving ample opportunity to destroy most weeds—but at considerable cost. The deeper cultivator and rotary treatments were made twice, as the full depth could not be achieved in one operation. The shallow plots had only one cultivation. As pointed out already, a single cultivation serves only to germinate the weeds—and so would a double cultivation carried out without a suitable period between operations, though no doubt some weeds would be too deeply buried to germinate by the deeper tillage.

It should be stressed that it is part of the organic method to use one light cultivation to germinate the weeds and a second subsequently, so that their remains provide food for the crop, not competition. If need be, a third cultivation may be given—one is going to lose very much less by a further week's delay than from weed competition. This Rothamsted experiment contains bad farming practice and has not, therefore, the value it might have.

Furthermore, Rothamsted soil will be much lower in readily available organic matter than the soil of an organic farm. Consequently, the ill effects of deep cultivations will be offset by the improved aeration.

It is interesting to learn that the N.A.A.S. have recently set up an experimental trial in cultivation methods. They have a number of combinations of deep and shallow cultivation, using horses, wheeled tractors, track-laying tractors and rotary cultivators. The manuring, however, is of the orthodox kind, and it seems likely that the full benefit of shallow cultivation will not be forthcoming. Announcing this experiment at a horticultural conference recently, a senior officer of the N.A.A.S., in reply to a question, said that they would not be

[1] E. W. Russell, Rothamsted Expt. Sta. Rep., 1949, 138.

Plate 17*a*

AVAILABILITY OF POTASH IN A SOIL ENRICHED WITH ORGANIC MATTER FROM A GREEN MANURE CROP (SWEET CLOVER)

Potatoes 17th June, Var. Home Guard giving $1\frac{1}{4}$ lb per root. Final crop 10 tons per acre at the end of June.

Left. Sweet Clover only.

Right. The same with 2 oz. per sq. yd. of sulphate of potash added.

Potash analysis:

Before green crop 0·001 per cent. End of March.

After green crop 0·006 per cent. End of March.

In the potato crop 0·009 per cent. Mid June.

In vacant plot 0·011 per cent. Mid June.

Plate 17*b*

EFFECT OF ORGANIC MATTER (COMPOSTS) ON PLANT GROWTH IN POOR SOILS

Boxes on the left—neutral soil (Weston Common).
Boxes on the right—acid soil (Wareham Heath).
Left half of each box: Trifolium pratense.
Right half of each box: Trifolium hybridum.
Compost added to the two boxes at the rear.

(By permission of Dr. I. Levisohn)

(a) Showing 4–5 feet of growth by early June.

Plate 18

SWEET CLOVER

(*Melilotus Alba*)

(b) Litter remaining the following spring after autumn rolling.

applying 4 to 5 in. of compost as they just had not got it. It is a pity that such views have taken hold—15 to 30 tons to the acre, according to the type of cropping used, is sufficient.

A criticism often levelled against rotary cultivation is that it damages soil texture by pulverizing the soil too finely, leading to caking in wet weather. Also that it destroys worms. If shallow cultivations are made and organic matter is adequate neither occurs. I have found that even if the soil is cultivated in a rather wet condition, and the surface crumbs are rather pasty, their crumbly condition recovers in a week or two, if conditions are suitable for the activity of micro-organisms. Provided there is a good supply of organic matter and it is not too cold or dry, they will quickly penetrate the lumps and aerate them. Or if panned by rain the surface does not set hard but remains crumbly and perforated by worm-holes. I have seldom seen a worm brought to the surface by shallow rotary cultivation, and suspect that the intense vibration of these machines sends them to the bottom of their burrows.

It may also be said that a clean seed-bed cannot be obtained with a rotary cultivator used for shallow cultivations. It can if used often enough, and it may be used at high speed two or three times for the same expense and time as one ploughing. I have used no other machine for seven years, with a rainfall of 45 in., starting with a field solid with couch, coltsfoot, bindweed and silverweed.

The sowing of crops is greatly simplified under shallow cultivations. Instead of endeavouring to produce a fine tilth from what is in effect subsoil or something near it, and inevitably containing more of the finer clayey constituents washed down from the surface, one already has a topsoil containing more of the coarser particles and a relatively high concentration of organic matter. This lends itself to easy cultivation. In addition, the ready-made drainage system of undisturbed soil carries surplus water away quickly and yet at the same time the soil below the tilth, being continuous with the subsoil, will remain moist. No longer does one have to contend with a dry, knobbly rubble or plastic dough. One can sow one's seeds in the firm assurance that they will germinate whatever the weather that follows, provided the soil below has not been dried out by a preceding crop or weeds. Certainly spring sowing is assured.

It follows, too, that seed can be sown much more thinly than usual, saving labour and expense both in seed and thinning, and as weed competition is slight each plant gets a better start, growing on without check.

K   145

Most implements, of course, are not adapted to this system of cultivation, but there are now a number of rotary cultivators on the market, and light track-laying tractors—a light tractor is all that is required, as there is no heavy draught work to do. I would stress the value of the track-laying tractor—the ordinary heavy wheeled tractor is an abomination to the organic farmer—the worse because the damage it does is unseen.

It is worth mentioning in this connection that Mr. A. W. Secrett, whose large market-gardening firm enjoys a high reputation for quality, was worried a few years ago by an increase in waterlogging during winter in a light and apparently well-drained soil. He was advised by Mr. Basil Furneaux, a leading soil consultant, that the trouble was due to a pan at plough depth. Mr. Secrett immediately scrapped his wheeled tractors and bought track-layers, with the required result.

For the market gardener, at any rate, the narrow-width rotary machines are ideal for inter-row cultivations. Seed drills are not always satisfactory unless preceded by a disc coulter to cut the litter which sometimes occurs. Generally speaking, until more suitable implements are manufactured, it is advisable not to sow small seeds directly after a crop producing heavy litter. Potatoes or cabbage (planted) are more suitable propositions. Broadcast corn would be satisfactory disced or harrowed in, as would beans or peas. The finer seeds are best left until the following season—kale, turnips, carrots and the like.

Regarding the matter of weeds, it is asserted by some (F. C. King, *The Weed Problem*) that these have their uses. In moderation this is probably true. As shown before, provided the soil is fertile, it does not always make any difference whether the crop is weeded or not. This obviously cannot be applied to small growing plants, as some horticultural crops, and young seedlings. If the crop is going to be literally smothered, then weeds must be removed. In most farm crops growth is usually strong enough to overcome the weeds by mid-summer, and their decaying remains will add something to the crop in August and September. Another point, too, is that every soil has its particular weed flora, dependent to a large extent on the nutritional status of the soil. If the soil is deficient of a particular nutrient, say phosphate, those weeds that thrive are obviously rather better than most at extracting phosphate, and when they die the crop will acquire the phosphate extracted—providing the life of the crop is extended beyond that of the weed. It cannot be said that weeds will be of any value in this respect for quick-growing horticultural crops such as lettuce or radish. Rather the reverse, in fact, though F. C. King suggests the possibility of symbiotic relationship. I do not know of any concrete examples apart from leguminous weeds, but note a later reference to possible benefits in respect of virus diseases.

There is one time of year when weeds have an undoubted value, that is in late autumn and winter. They hold plant nutrients against winter losses, and when cultivated into the ground in spring provide the seedling with a useful send-off. It is very noticeable how much faster seedlings grow if the ground has a green cover in winter than when sown on ground left bare.

This, of course, applies to annual weeds. While perennial weeds undoubtedly bring up material from the subsoil, such weeds in arable land have far too big an advantage over seedling crops in the way of stored reserves, and for this reason alone cannot be tolerated.

In practice most of us, unwittingly, use our weeds to the best advantage. They are hoed off, they decay on the spot, releasing what value they have, until the crop is of such a size that it can look after itself. Some farmers make use of them as winter fodder, and claim that they have a value not shown by grass at this time. Not unlikely, as most of the nourishment of grass is stored during winter at the base of the leaves and stems or even underground, while those weeds growing in winter have most of their nourishment well above ground, in their leaves and stems.

While on the subject of weeds, it is probable that the large quantities of weed-killers sometimes used on our crops have a detrimental effect on the soil population. It seems likely also, however, that some kinds, used in moderation on weeds in the seedling stage or even larger, would be relatively harmless. In particular, the hormone class of weed-killers, which are in effect organic materials used in much higher concentrations than are normally found in the plant, thus upsetting its metabolism. As these materials are found naturally they must be decomposable, and would therefore only present a temporary and probably slight distirbance of soil functions.

It is known, for instance, that b-indolyl-acetic acid and a-napthyl-acetic acid are rapidly decomposed by soil organisms, while the most commonly used hormone weed-killer, 2:4: Dichlorophenoxyacetic acid (2:4:D for short) is relatively resistant. Even this evidently disappears in a few weeks, or indeed much of our agricultural land would already be sterile. Even so, a crop so weedy that it requires cleaning by weed-killer is evidence of bad husbandry—but emergencies do arise sometimes.

Another weed-killer which has a place under certain circumstances is TVO (tractor vapourizing oil). It saves an enormous amount of work with carrots and allied crops, and as it evaporates within a few hours, little harm is done. I am thinking in terms of 5 to 10 gallons to the acre, not 50 or 80, as commonly recommended, directed over the drills directly after emergence of the carrots, when the surface of the ground is dry.

In very heavy or waterlogged soils, and they are almost synonymous in winter, ploughing may be advisable. Such soils are very slow to warm up in spring owing to their wet and dense nature conducting heat to the wet subsoil rapidly and even more to the constant evaporation at the surface. Cutting off the topsoil from the subsoil and introducing air by ploughing will check these processes and permit the plant to make an earlier start than would be possible under shallow cultivations. It should be added that all subsequent cultivations and machinery should be of the lightest so as not to restore contact with the subsoil by compaction. Ideally, such soils should be avoided for arable work unless drained, as panning is inevitable, but it is not always economically possible.

Waterlogging has another effect which is more easily dispersed by ploughing: that of the formation of toxic materials, particularly hydrogen sulphide. The access of air stops the production of these substances and rapidly dissipates them.

The management of these soils is difficult under any system of cultivation; it will, therefore, be helpful to discuss the means by which natural vegetation is enabled to survive on such soils.

Such land is invariably covered with coarse grass and rushes which form a fibrous surface mat of roots and dead herbage. These roots largely die off in winter and are renewed each summer. Thus a spongy surface layer is developed which becomes well aerated as the water level falls in spring. Plants are thus able to exist without the aerating influence of worms which cannot live under such conditions.

When such land is converted to arable crops this spongy layer is broken up and the organic material is oxidized by bacterial action and not replaced, as arable crops do not leave any great quantity of fibre in the soil. Air will be unable to penetrate the surface, with fatal results to most plant roots except perhaps to those adapted to conditions of poor aeration, or those of annual plants which, though normally penetrating to deep and relatively airless regions, can so change their habit as to keep near the surface.

With the absence of worms it seems that any organic matter applied to the soil will have to be mechanically incorporated, otherwise it will lie as an inert mass on the surface, unable to decompose until the damp conditions of late summer or autumn, when it will be too late to be of value to the plant. Alternatively, such land must be laid down to grass at frequent and lengthy intervals.

Similar effects to these may occur on better-class soils during prolonged, cold, wet weather. Loss of nitrogenous materials also occurs, partly by direct leaching and also by the death of the microbial population and the loss of their more soluble constituents.

An interesting example of these difficulties as seen on the 'no digging' experimental plots of Mr. J. L. H. Chase at Chertsey after the wet winter of 1950–1. Many crops the following summer were poor. These plots are on a heavy clay soil with little slope and a considerable degree of waterlogging was inevitable. In districts of heavy rainfall the use of such soils for horticultural crops would be impossible.

Even potentially good soils need careful handling in such seasons. Mr. D. H. Pizer, N.A.A.S., suggested at a Chelmsford conference in 1951 that the collapse of soils due to heavy machinery and low organic content was the cause of some apparent mineral deficiencies in crops.

The handling of leys, apart from manuring, to provide the maximum yield, is a controversial subject on which the clarification of the basic principles involved will be helpful.

A ley, especially of the type used by organic farmers, is a mixture of species of different sizes and rates of growth. Left to their own devices a few species would become dominant.

This is prevented by grazing or mowing to reduce all plants to the same level and allow them equal opportunities to develop again. There is, however, a danger in cutting or grazing too hard. When a plant is cut close to the ground its food supply, which is manufactured in the leaves, is cut off and much of its root system dies as a result. Unless there is a reserve of nourishment (carbohydrate) in the roots and stem to furnish a new shoot, which in turn must nourish fresh roots, the plant will die. As a rule there is sufficient for this purpose. It is obvious, however, that the more leaf and stem remaining to the plant the more rapidly will it be enabled to recover. It is enabled to start into growth again quickly. Anyone who has noticed how the irregularities of scythed grass are quickly accentuated will appreciate this point. The farmer who, from a sense of tidiness or mere parsimony, cuts or grazes his crops bare is defeating his own object.

Figures demonstrating this effect on lucerne are shown below.[1] Cutting was for three months—April 13th to July 6th. Weights are in grammes of dry matter left by the end of the experiment, with eight plants per treatment.

Treatment	*Cutting height i.e. length left on the plant*					
	1 in.	*3 in.*	*6 in.*	*9 in.*	*12 in.*	*Control*
Weekly	29·03	63·07	78·14	85·01	99·59	79·53
Bi-weekly	45·39	64·42	91·19	96·33	102·40	79·53
Monthly	78·91	83·73	99·78	113·19	84·44	79·53

Amount of renewed growth above the cutting level for the first two months. (Much of the renewal growth sprang from the root crown.)

	1 in.	*3 in.*	*6 in.*	*9 in.*	*12 in.*
Weekly	7·75	14·66	16·05	12·55	9·86
Bi-weekly	15·12	16·56	20·63	17·86	13·93
Monthly	29·07	27·10	26·44	25·25	12·00

These figures suggest that when cutting for hay it is best to cut not more often than once a month, to a height of 1 to 3 in., but that haying does not make use of anything like the plant's full potentialities. If permitted to keep from 6 in. to a foot of stem, large numbers of basal shoots develop, which give two or three times the crop. Thus, it seems that for the most efficient use, it should be grazed at intervals of two to four weeks, but without permitting it to be eaten down to less than 6 in. Also, the more frequent the grazing the lighter it must be, and conversely.

The amount of root growth is also related to the top growth (weight in grammes).

---

[1] S. C. Hildebrand and C. M. Harrison, *J. Amer. Soc. Agron.*, 1939, 31, 790.

Cutting Interval	Cutting Length					
	1 in.	*3 in.*	*6 in.*	*9 in.*	*12 in.*	*Control*
Weekly	3·22	8·76	11·46	22·31	39·39	41·52
Bi-weekly	5·21	13·21	17·14	24·16	30·64	41·52
Monthly	11·09	18·24	20·00	33·78	38·88	41·52

The amount of root development, and incidentally storage capacity for the following season, is inversely proportional to the amount of cutting. The roots of the most closely cut plants were only one-twelfth of the uncut, and even when cut to the ground monthly only about a quarter of the root development occurred. Thus, some of the lucerne's great advantages—its deep rooting and mineral extracting capabilities—are barely utilized.

Another interesting experiment showing this principle was done by Sir Albert Howard on the indigo plant. Instead of cutting the plant to the ground, as was the usual practice, he left a branch on each plant to grow. This reduced the first cut by about a third. The second cut, however, was four times as great on those plants left with a branch as those cut to the ground. The total yield for the two cuts showed an increase of about 30 per cent in favour of leaving a branch.

These effects are applicable to all plants, though their severity will depend to some extent on the natural height of the plant. Thus, a plant naturally growing 6 in. high would not suffer nearly so much when cut to 3 in. as would one growing 12 in. high. This, in fact, is the means by which we are able to maintain a balance in a mixed ley.

It seems, then, that the most profitable method to handle grass is to leave 2 or 3 in. after each grazing, then allow it to grow until a further grazing is required—the off and on system—but taking care not to graze bare. And the larger the top growth the deeper the roots will penetrate to comb the subsoil.[1]

Providing grass can be got off the ground quickly, as for silage or drying, it would prove profitable to leave a few inches of grass below the mower. The rapidly growing aftermath would cause complications in the case of hay-making if delayed, so such procedure would

[1] E. Klapp, *Pflanzenbar*, 1943, 19, 221.

be best avoided for this purpose, especially in wet districts, unless tripodded.

The urge to extract the last ounce of reward for our labours is common to most of us. But it has its dangers. The removal of every possible scrap of readily decomposable organic matter is to an increasing extent each year jeopardizing the succeeding crop. Even, in my own experience, leaving the tops of carrots to rot rather than feed them to animals, has given the seedlings of the following crop a surprising lead over those growing on bare land—at least under shallow cultivations. It is likely that, given this lead, one receives back several times more than was lost the previous autumn, as the growth of a plant is cumulative.

It is interesting to note that Edward Faulkner, in his book *Soil Restoration*, claims that he has restored his land in seven years to a high degree of fertility by the simple expedient of allowing crop waste to rot on the surface.

Deep-rooted leys should be self-sufficient in their manurial requirements, but there is often on well-managed farms a quantity of urine to be disposed of. This may be spread on a ley but only in small quantities. It has been found that the addition of a nitrogenous manure to grass in early spring will often lead to a shortage of grass later in the season.[1]

The reason given for this is that the large amount of grass roots which die off when the grass is cut take up any nitrogen in the soil for their decomposition. Doubtless this plays a part, but it would occur whether spring nitrogen was added or not. My own opinion is that the decaying materials in the soil, which should have been providing nitrogen later in the season, were hastened in their destruction by the use of the early nitrogen. It had thus released its nitrogen when there was already plenty for the crop, by the mechanism described in Chapter V, so that it was lost to the atmosphere. Nitrogen release is at its maximum in late spring and early summer, so that a soil requiring a nitrogenous top dressing at this time must be in exceedingly poor heart.

There does not seem to be the same danger in applying moderate amounts in, say, August or September after the hay is cut, when some dead grass roots and stems will have accumulated. Dung also may be applied at this time, but again there is no point in applying too much, as it is quickly lost. On theoretical grounds only 1–2 tons per acre, or 200–400 gallons of urine, are required for every 1–2 tons of dead

[1] Rothamsted Expt. Sta. Rep., 1951, 40.

fibre, provided by the grass and roots. It is a practical impossibility to spread such small quantities, so that in practice the minimum quantity compatible with even distribution is used. It is, unfortunately, relatively costly to spread small quantities, so the necessity is best avoided if possible by keeping cattle on the pasture as much as possible.

From a theoretical standpoint, nitrogen should be an embarrassment on a well-managed farm when one considers the 100 or 200 lb. fixed per acre by legumes. An impossibly large quantity of produce would have to be sent off each acre to carry this away; crops normally only carry off amounts varying from 10 to 50 lb., and much of this is fed back. Losses are inevitable, but provided reasonable precautions are taken there is no need to be concerned about them.

There is a crop of increasing interest in this country, sweet clover, which is of particular value to the stockless arable farmer. A more detailed consideration of this crop will be found in the chapter on market gardening. It provides the soil with a much larger quantity of organic matter than any other crop and, in addition, a very large amount of fixed nitrogen, with minerals extracted from the subsoil. It is particularly valuable on dry, thin, and light chalky lands, often growing to 7 or 8 ft. in its second year.

As a crop for fodder, it will produce four or five heavy cuts per year of high-protein food (leaf contains 6·6 per cent protein). The plant must be cut for silage or grazed when not more than a foot high, as it soon becomes woody. It is not advisable to attempt to make hay, as it is thick-stemmed and slow to dry. There is another reason. The plant, unfortunately, contains a bitter principle, coumarin, which makes the plant rather distasteful to animals at first. It also has the power of preventing the blood from clotting, but this is greatly accentuated in hay that has become mouldy. It may be mentioned that a coumarin-free form has been developed by the Canadian Government Research Farms.

It is often sown with a cereal or grasses so that a crop is obtained in the first year. In wet districts its use in a cereal is not advisable, as it makes rapid growth late in the season and may complicate harvesting.

In view of the plant's remarkable powers of restoring soil fertility, it is surprising that it has not been used to reclaim our open-cast mining sites. Dr. R. M. Salter, Chief of the Bureau of Plant Industry, U.S.A., claims to have cropped 59 bushels of oats per acre where the topsoil had been bulldozed off. Lime, potash and phosphates were applied to the subsoil and a crop of sweet clover grown for only a year. A little artificial nitrogen in the form of ammonium phosphate would be beneficial in a case like this. Even leguminous plants require a start in life before they can begin to fix their own nitrogen. Organic nitrogen would not be an economic proposition in such circumstances.

Many crop plants are particularly good at extracting minerals. Phosphates, for instance, can be extracted from relatively insoluble compounds by lupins, lucerne, sweet clover, swedes, turnips, mustard, buckwheat and millet. Grassland, on the other hand, may suffer from slight phosphate deficiency without showing it, though the grazing animals may suffer. This is a subject on which research work is needed; there is little reliable information to be had.

The reseeding of mountain grazing is a matter of importance to food production. It may, however, have a danger. Under the heavy rainfall of these areas, provided drainage is good, organic matter is rather quickly used up by micro-organisms—but is replaced by the tough grasses and plants which are rather inedible to animals. Thus a fairly stable ecological situation has been set up. The immediate result of ploughing up and liming (and even more of adding nitrogen to) the thin stony soils of mountain land will be a fairly rapid breakdown of organic matter and release of nitrogen. This, in turn, will lead to good growth in the new pasture—for a time at least. When a new palatable ley is put down, the herbage matter will be consumed by the animal and, furthermore, the dung and decayed plant remains will be of a softer and more easily decomposed nature, leaving little organic matter for the soil, lacking the more lignified components of the original herbage. It is possible that unless the seed mixture is chosen carefully to contain some of the tougher grasses, and management controlled to prevent close grazing, a deterioration will set in, leaving the ground considerably less productive than the original herbage.

If the improvement in herbage is brought about simply by drainage, however, circumstances are much more favourable for lengthy improvement. Such soils are generally high in organic matter for one thing, and the original herbage may be improved sufficiently by the elimination of excess water and rushes.

It should be remembered, too, that the introduced species may not be so efficient at extracting minerals under poor conditions as the original herbage. It has been found, for instance, that sheep have done better on such grazing than on the improved ley.[1,2]

[1] B. Thomas, *et al.*, *Emp. J. Expt. Agric.*, 1945, 13, 93.
[2] J. N. Peart, *J. Brit. Grassl. Soc.*, 1951, 6, 219.

Subsoiling is a practice to be used with discretion. Its chief value seems to be to remedy the ills of ploughing, to break up any hardness in the subsoil. Once done, however, providing one can maintain a good worm population and rotary cultivate with a low ground pressure, there does not seem to be any need to repeat the process. It seems any benefit gained by permitting freer root action would be counter-balanced by more rapid drying of the soil and oxidation of organic matter. In wet districts it might be used with benefit, especially if run up and down the slope, to promote better drainage. But, as far as silt soils are concerned, they are best left alone, as far as possible. The fine aeration and drainage cracks in such soils are easily damaged and difficult to reinstate. However, as in most farming practices, a thorough knowledge of the soil and rooting habits of the crops is the final criterion.

VII

ORGANIC MARKET GARDENING

In the course of this book reasons have been given to support the view that there are sufficient food materials in the soil to support our crops without importation of fertilizer material from outside sources. Broadly speaking, this seems to be true. We have, however, the exceptional case of the intensive cropping of market gardens. Farm crops, generally, only remove a few dozen pounds of minerals per acre annually. Market-garden crops may remove a few hundred pounds and need their nutrients in a form easily available to them.

Generally speaking, market-garden crops remove far more minerals than farm crops and, what is more, they are sold off the land. It seems likely that few soils could stand this rate of cropping indefinitely. There are two courses open here. One is to mix market-garden crops with farm crops. This can quite feasibly be done with the bulky crops, such as cabbage and potatoes and is, from a national viewpoint, the idealized system. The other course is to import plant food into the market garden, and, under our present economic arrangements, is the only course for many. This fertility can come from other farms or areas in the shape of dung, straw, hedge-side wastes, etc., where the extraction of plant minerals is normally low.

A further development of this principle, whereby a market garden may become self-sufficient, is to put down a portion of the land to a deep-rooted mineral-extracting crop giving a large bulk of carbonaceous matter and nitrogen and either cultivate it in or transport it where it is needed. I use the former method on part of my own market garden. The latter is more suitable for fruit growers and an example is described in *The Grower* of 27th June 1953 (page 1236), where about 40 acres of lucerne are transported annually to mulch 35 acres of fruit trees. It should be remembered that the more intensively a market-garden soil is cropped, or the more it is manured, the more it

156

becomes dependent for its organic matter on an increasing area of outside land. A market gardener, therefore, may buy his dung, straw, compost, etc., from outside sources, or produce it from his own land, whichever is most economical. He may feel that producing it from his own land is reducing his production per acre. This is not so—market-garden crops are almost invariably grown on the fertility withdrawn from many more acres than the land actually occupied by the crop. To this extent, then, crop yields per acre for market gardens are meaningless.

QUANTITIES OF MATERIALS TAKEN UP BY VARIOUS CROPS
FROM AN ACRE OF SOIL

	Total Dry Weight of Crop in lb.	*N.*	*P.*	*K.*
Wheat, 30 bushels	1,530	34	6·2	7·7
Straw	2,650	16	3·0	16·2
Total	4,180	50	9·2	23·9
Beans, seed, 30 bushels	1,610	78	10·0	20·2
Straw	1,850	29	2·7	35·5
Total	3.460	107	12·7	55·7
Mangolds, roots, 22 tons	5,910	98	15·4	185
Leaf	1.650	51	7·6	64
Total	7,560	149	23·0	249
Hay, 1½ tons	2,820	49	5·4	42·3
Turnips, roots, 17 tons	3,130	61	9·8	90·1
Leaves	1,530	49	4·7	33·4
Total	4,660	110	14·5	123·5
Meat 250 lb. live weight (not fat)	—	60	4·5	5·0
Potatoes, 8 tons	4,480	61	28	100
Cabbage, 20 tons	—	195	65	220
Carrots, 15 tons (including tops, 8 tons)	—	115	40	156
Onions, 6 tons	—	24	12	26
Tomatoes, 40 tons (glasshouse)	—	134	42	215
Apples, 10 tons	—	13	6	30

In all cases the total amount of minerals used remains the same. Or we can import our minerals from deposits underground either in the natural state or as manufactured products, usually in the form of sulphates. For reasons already given, the natural product seems preferable.

We are, however, likely to run into difficulties by importing straight phosphate and potash minerals, especially as the manufactured product. We are replacing those minerals used in large quantities, but not the minor elements, unless they occur as impurities in the imported mineral. Difficulties also occur through mineral imbalance, an excess of one mineral making it difficult for the plant to take up another. Therefore, the importation of composts, dung and wastes from other areas seems preferable, and composted sewage wastes would conveniently fill a gap here.

Notwithstanding all that has been said about the difficulties associated with the use of artificial manures, there does not seem to be any intrinsic reason why minerals, apart from nitrogen, should not be applied to the soil, provided they can be applied in a form that has no side effects on it or the plant. Such a form would be a relatively insoluble compound but capable of being dissolved gradually by soil micro-organisms, without leaving any toxic or acidifying residues. Even so, providing there is plenty of organic matter and lime present, moderate amounts of mineral sulphates may cause no harm. Furthermore, no more should be added than is required for the crop or rotation of crops, rather less in fact, to avoid a build up and possible imbalance. Under conditions of ample readily available organic matter the fixation of minerals by the soil does not seem to occur. There is already a useful phosphate compound, phosphate rock; but, unfortunately, there is no comparably rich source of potash apart from the rapidly soluble potash salts. There is an opening here for the fertilizer manufacturers to produce such a potash compound. The only solution at present is to add to the soil potash salts, or the sulphate, ready mixed with the equivalent amount of calcium, as lime to replace that which will be lost when the acid radicle combines with it, and well before the crop is sown. By this means, the soil will remain at the same level of acidity throughout the growth of the crop.

The only remaining danger to the use of minerals direct—especially as purified products—is that one is not replacing trace elements. The chemist can help us a good deal here by analysis of the total constituents of the soil—not merely the 'available' ones. Unfortunately, our knowledge of the plant's needs in this sphere is, as yet, very incomplete, so those who are compelled to maintain their cropping in this way must take the risk—admittedly slight these days—of undetected shortages.

It must be borne in mind by extremists of the organic 'school' that in many industrialized areas there are quite large areas of very intensively worked horticultural land, especially glasshouses, where the acquisition of sufficient minerals in an organic form seems to be an economic impossibility at present. Perhaps when the composting of sewage wastes becomes more general the position will be eased. It is a situation particularly demanding of research. It must be remembered, too, that though such produce may well be deficient in some respects of vitamins and minerals necessary to human health (see Chapter X), they form a comparatively small proportion of our diet —though unfortunately a large proportion of our raw salads in winter time.

Quantities of compost for market-garden crops are not excessive. A normal dressing of farmyard manure is 60 tons per acre for market-garden crops, whereas only 30 tons of compost are needed, rising to 50–60 tons, if possible, in heated glasshouses, which generally carry two or three heavy crops in a season. If such dressings can be given, there is no need for additional minerals.

Potting composts may be made up with compost instead of the more usual peat, but it needs to be very well rotted or nitrogen starvation may ensue. A seed compost may be made of 3 parts medium loam as weed-free as possible, 2 parts of compost and 1 of coarse sand or grit, passed through a $\frac{3}{8}$-in. sieve. One must, of course, ensure that the degree of acidity is within the range acceptable to the plant by adding ground chalk or limestone, if necessary. If the plant is to stand for several months in the pot, then 2 parts of compost to 2 of loam and 1 of sand should be used. If the plant is of vigorous growth, as in the case of tomatoes, cucumbers and chrysanthemums, some additional feeding—with well-rotted compost and perhaps dried blood or hoof and horn—will be needed.

However, in the sphere of carbohydrate material and nitrogen the general market gardener, and perhaps even the glasshouse man, can help himself to some extent. Both these are derived ultimately from the atmosphere, which is inexhaustible.

It is appropriate at this point to introduce to the market gardener, particularly the extensive grower, the plant sweet clover (*Melilotus alba*) which, I believe, can play a very important part in maintaining the fertility of market-garden land.

Most market gardeners, and indeed farmers, are wedded to the concept of ever-increasing yields per acre as a means of competing with rising overheads. Unfortunately, the law of diminishing returns becomes increasingly operative.

As labour and the running of machinery are by far the largest items of expenditure, it might be profitable to ignore the yield per acre at the present time and concentrate on yield per man and reducing cultivation costs. He might find it more profitable to retain only his best workers, cut down his cropping and lay down a third or quarter of his land to a green-manurial crop which will provide sufficient fertility to make him largely self-supporting in organic matter. The smaller amount of produce (and of better quality) finding its way to the market would not be without effect on market prices also. Certainly, in my own experience, the use of this crop has proved a boon which I can hardly measure. My import of farmyard manure has dropped to a mere 2 or 3 tons per acre reserved entirely for making compost—the greater part of the land is having its fertility built up by the use of sweet clover alone.

An account of the gradual release of potash by this plant has already been given. For land short of phosphate it is a good plan to manure the land before sowing with rock phosphate which it renders available for the next crop. I regularly obtain crops of between 10 and 20 tons per acre of Majestic potatoes with no other manure—this in spite of the fact that they are cut down with blight in this district by the end of July. It has been found, too, that the addition of compost at 15 tons per acre gives no further increase in crop.

Other crops, especially strawberries and root crops, do very well after it. Cabbage, lettuce and similar greedy crops may need a little nitrogenous manure as well, if the ground was originally poor, though it is not essential if quick results are not required. Cloched tomatoes this season, 1953 (Amateur), produced a total of 7 cwt. of fruit per 100 yds. of run with no other manure or top dressing. Approximately $5\frac{1}{2}$ lb. per plant were given by the end of August, even though the plants were sown and raised without heat.

This plant, in addition to extracting minerals, including calcium and magnesium, is deep rooted, drought resistant, and will penetrate any plough pan, providing the land is not waterlogged. It will provide during its second summer, if allowed to mature, 10 tons of dry organic matter per acre in the tops alone, which is equivalent to 40 tons of dung, and containing nitrogen, which I can only provisionally estimate at the equivalent of 30 cwt. of sulphate of ammonia, but

SWEET CLOVER (*Melilotus Alba*)

(*a*) Sweet clover and lettuce intercropped with carrots. Mid-July, lettuce ready for cutting.

(*b*) Incorporation of the litter in preparation for a potato crop. Right, one light rotary cultivation; *Left,* a second cultivation to 3 inches.

(*a*) Plum. Victoria, five years old, grown in lucerne and weed grasses, uncut, with adequate rainfall. Three feet of extension growth in spite of a very heavy crop requiring staking.

Plate 20

**FRUIT TREES
UNDER
ORGANIC
METHODS**

(*b*) Apple. Rival, six years old under similar conditions, showing a heavy crop and ample extension growth. Some wind damage to the leaders.

mostly in a gradually available form. There are also the root systems with their manurial content to be considered.

Being a biennial plant it occupies the land for the whole or part of two summers. It can, however, be sown as late as the first week of August, if necessary, and intersown between early harvested crops, so that it occupies the ground for only one summer. It may also be used as an annual, sown in April or May, as a nitrogenous manure for the next season's crop, with good effect. In this case only 1 or 2 tons of organic matter will be formed but relatively much higher in nitrogen. I have taken 14 tons of potatoes to the acre after such treatment. This is double the yield for the county. It might be more profitable on this basis to crop the land on alternate years rather than every year—the same crop for a very considerably reduced cost of cultivation.

If poor land is being reclaimed, i.e. poor in the sense that it is low in organic matter, it is best to sow fairly early the previous summer, not later than June, to get it well established the first season. It is also advisable to inoculate the seed with the necessary nodule bacteria— this gives it a better start, unless allied plants—lucerne and trefoil or medick (*Medicago lupulina*)—are, or have been, plentiful in the field. It is sown at the rate of 15 lb. to the acre, preferably drilled, if inoculated, or broadcast, if it or allied plants have been grown before.

It may be sown between carrots, turnips or beet for bunching if the crop is drilled at about 18 in., and allowing the crop about three-weeks' start. This also allows ample time to clean up any young weeds that have developed between the rows. It is useful, too, between brassica seedlings, radish and lettuce—in these cases it may be drilled at the same time between the rows or mixed with the crop seed. I find a triple combination of lettuce and sweet clover, sown mixed at 18 in., with carrots sown between three weeks later, a useful combination where land is at a premium. After cutting the lettuce it is necessary to cut back the clover hard to prevent competition with the carrots. They will appreciate the extra nitrogen provided by the clover.

When the crop has matured, about August, it may be cut down with a rotary cultivator or disc. A second cultivation will be needed a fortnight later when the material has rotted somewhat. The ground will then be in ideal condition for planting spring cabbage or straw-berries.

Alternatively, it may be rolled, in late autumn, leaving it then until spring, when the stems will have become brittle and easily disin-tegrated by rotary cultivation or discing. The large amount of litter, however, makes it unsuitable for small seeds; potatoes or transplanted crops are more suitable.

Regarding market-garden cultivations, the broad principles are the same as for farming, i.e. maintaining a soil cover throughout the year and shallow cultivation of the soil. Usually the cover will consist of a cash crop, but occasions do occur, especially on the extensive market garden where land may be idle for a short period. In this case it is hard to improve on a crop of mustard—there does not, unfortunately, appear to be a nitrogen-fixing crop with such comparably fast growth. It is a useful autumn crop as it may be left to hold nutrients from the autumn rains and provide readily available food for soil organisms, especially worms. During the winter it dies off, leaving clean ground for the spring crops. The folly of ploughing such crops in, in autumn, as is commonly done, has already been enlarged upon.

There are now a number of two-wheeled rotary cultivators designed for the market gardener. Unfortunately, they are usually far too robust and heavy for surface cultivation. Mr. E. R. Hoare, director of the horticultural section of the National Institute of Agricultural Engineering, said recently that number and depth of cultivations made little difference to yield but that pressure did—even on light soils in dry seasons. It is one of the difficulties of organic farming that the soil is very retentive of moisture and therefore soft, and yet remains capable of being worked at the surface. It is for this reason that the disc is not entirely satisfactory; there is a risk of a pan forming close to the surface.

The cultivator, therefore, should have broad wheels and not weigh more than 2 cwt. or so. Two machines of particular value are the Clifford Rotary Cultivator (Mk. A.1) and the Howard Bantam. That is not to say there are no others, but my personal experience is limited to these two. For large-scale work there are the Bristol and Ransomes M.G. tractors, both of the caterpillar type.

One method of avoiding soil compaction in a market-garden crop is to use a wide wheel-based tractor, omitting plant rows where the wheels run, leaving permanent tracks, through the life of the crop— or permanently, if desired. As wheels do not have to pass through the remaining rows the crop can be spaced closer, giving the same plant as before. These wheel-tracks are also useful for harvesting the crop and avoid treading between the crops. Useful, too, for throwing any larger weeds, if hand-pulling should prove necessary, where they can be dealt with by a narrow-width rotary cultivator.

The permanent bed system for crops, rather than a square plant, is much to be preferred for the same reasons and is a development of the foregoing. Where land is not at a premium, 4-ft. beds may alternate between 3-ft. pathways, or even 4-ft. paths where cloches are used. They may, however, be reduced to 2 ft. if necessary. The land is not wasted so much as might appear at first sight, as crops and their roots often spread out for a considerable distance over the path if planted to the outside of the bed. In any case, the improvement in growth due to a well-aerated soil is ample compensation, especially in wet areas or heavy soils. Every footprint seals the surface for a time. The system is of particular value for permanent or semi-permanent crops, strawberries, asparagus, flowers, and things requiring attention in winter time, anemones and bulbs. All hand cultivations become much easier in the soft soil.

Weeds generally are a much more serious problem to the market gardener than the farmer. The main difference, in fact, between farm and garden crops is that the latter need more cultivation, they are not so robust and able to compete with weeds. The methods outlined for preparing a seed-bed for farm crops are, however, equally applicable: a series of light cultivations to germinate and kill the weeds, rather than to plough them in for next year. Thus, one gradually eliminates the long-lived seeds from the topsoil and one is left with those types which germinate readily, generally those weeds which mature quickly, such as groundsel and annual meadow grass. Chickweed only seems to thrive well where there is much easily available nitrogen. However, their ready germination is their undoing, as they are thus more readily destroyed. One can also help matters by growing the more robust transplanted crops—brassicas—in the rotation.

There is a tendency, particularly when land is limited, to plant too closely. A good example is lettuce, often planted in glasshouses at 7 or 8 in. square or even less. Even with the smallest varieties this hardly gives room to develop a heart, and the risk of disease is extreme. My own lettuce are left at a foot square for the simple reason that they are usually a foot across. As a result, the price obtained is considerably increased, while disease risks and handling costs are equally reduced. Cheshunt Research Station discovered recently that 8,000 tomato plants to the acre bore the same crop as 16,000 plants, though, of course, there were a larger number of early trusses in the latter case. Whether this would compensate for easier working and freedom from disease is very hard to say without further experiment.

There are a number of difficulties associated with present-day methods of glasshouse management, not the least of which is the accumulation of salty residues from artificial fertilizers and to a limited extent from concentrated organic fertilizers. These are almost impossible to remove by flooding, as the water tends to run down soil cracks, by-passing the bulk of the subsoil. Possibly several weeks of fine mist-like spray might do it, but equipment and water for such a project is expensive. The issue is easily surmounted if the necessity for fertilizers is avoided. Composts generally are low in salty materials, with the possible exception of sewage composts, but even in this case exposure to rain for a time would wash out most of them. Where fertilizers cannot be avoided, those types containing little or no sulphates or chlorides should be chosen.

The need for frequent steaming is probably a reflection of unsuitable biological conditions, particularly a shortage of readily available organic matter which is very largely destroyed by bacterial action under the influence of nitrogenous fertilizers. The action of steaming on the soil has been discussed before—briefly, it kills the bulk of the soil micro-organisms and leaves their nitrogenous remains available to the plant, together with quantities of readily available phosphate and potash. (Bacterial protoplasm contains up to 30 per cent potash by dry weight.) Thus, very vigorous growth ensues at first. However, before the end of the season the micro-organisms have returned, including pathogenic fungi, which increase rapidly,[1] having little competition or antagonism from the other soil organisms, there being little readily available carbohydrate for them to feed on, so the steaming must be repeated. By adding energy food to the soil, with its biological content, a normal balance of micro-organisms is restored (providing the soil is otherwise suitable, i.e. not salty or perhaps saturated with fungicides). (See also page 199.)

The ease with which diseases can increase in a sterilized soil is convincingly demonstrated by the following table; where barley plants growing in sterilized soil are infected with a soil fungus, *Helminthosporum*. (From Christensen.)

NUMBERS OF STUNTED PLANTS

H. Added	H.+Soil Infusion Added
46	6
33	6
25	8
10	3

[1] S. A. Waksman, *Soil Microbiology*, 1952, p. 300.

Helminthosporium alone added to the soil multiplied rapidly, giving high numbers of infected plants, but when the normal soil organisms were introduced at the same time the competition was such that only a small proportion of the pathogenic fungus was able to infect the barley. It seems that unless every trace of disease is removed by steaming, it is worse than useless as a disease control, and under most conditions it is a practical impossibility to do this. It might, in fact, be wiser to sprinkle the sterilized soil with fresh, clean soil before planting the crop. Indeed, this has been found to be so in forest-tree nurseries.[1]

Of great practical importance to the commercial grower is the quality of his produce. A very difficult point to define—it depends what is meant by quality. Also, different people evaluate certain characteristics in different ways—some like a floury potato, some like a firm one. Nobody likes a black one. The appreciation of quality may even vary from time to time as a matter of fashion. Public taste appears to be changing from the crisp acid type of apple to the softer sweeter kinds.

From the commercial angle, the producer must offer something of attractive appearance, and preferably flavour, that will pack and travel without appreciable deterioration.

Very little work has been done on the subject, unfortunately, though the blackening of potatoes during cooking was considered by Rothamsted to be aggravated by the use of manures containing too much nitrogen in relation to phosphate and potash.

However, one very interesting experiment on the quality of apples in relation to manuring, by the Massachusetts Agricultural Experimental Station, was quoted in *The Grower* of 2nd May 1953, and some details are worth mentioning.

Some ten different treatments were used, per tree.
(a) 70 lb. hay mulch containing, on analysis, 0·67 lb. N., 0·10 lb. P., 0·75 lb. K., 0·20 lb. Ca, and 0·08 lb. Mg.
(b) Twice this quantity of hay.
(c) Three times as much.
(d) 70 lb. hay + 0·67 lb. nitrogen as ammonium nitrate.
(e) 2 lb. ammonium nitrate (35 per cent nitrogen).
(f) 4 lb. ammonium nitrate.
(g) 6 lb. ammonium nitrate.
(h) 1·33 lb. nitrogen in a complete fertilizer of 7–7–7 analysis.

---

[1] G. S. Perry, Penn. Dept. Forest and Waters, 1933, Ser. 4, No. 2.

(i) 2 lb. nitrogen in a complete fertilizer + 2 lb. ammonium nitrate.

(j) 1 lb. nitrogen as a urea spray on the foliage.

The results were measured as follows:

Treatment	Bushels per Tree	Percentage of Top Grade	Bushels of Top Grade per Tree	Firmness lb. Pressure
A	15·3	71·3	10·9	16·6
B	14·1	79·5	11·2	16·6
C	16·2	71·1	11·5	16·0
D	17·0	54·6	9·3	16·0
E	16·4	64·5	10·6	15·8
F	22·4	28·3	6·3	15·0
G	18·2	30·2	5·5	14·9
H	16·2	56·1	9·1	15·4
I	18·9	42·1	8·0	14·9
J	15·2	55·8	8·5	15·6

The treatments were carried over four years with trees of 'Macintosh' spaced at 27 per acre. The fancy grade is defined as 50 per cent of the apple surface of a red colour. Firmness is tested by the apples' resistance to bruising under pressure and it is also a measure of storage life.

It will be noted that the greater the amount of nitrogen used, the larger the crop—up to a point—and the poorer the quality. The heaviest nitrogen treatment (g) actually reduced the crop. (Incidentally, compare this quantity of ammonium nitrate, about 1½ cwt. per acre, with the quantities used by Dr. F. E. Bear in his experiments (p. 77, i.e. about 10 cwt., 20 cwt. and 100 cwt. per acre. We can only speculate what happened to the microbes, but we can have a good idea what would have happened to the crop.) Also that the biggest crops of top-grade apples came from trees fed organically. Note, too, the comparatively small quantities of nutrients supplied by treatment (a), about 20 lb. of potash, 18 lb. of nitrogen, and 2½ lb. of phosphorus per acre.

Other interesting observations were that the hay increased the availability of soil phosphorus and that an analysis of the trees' leaves showed a decrease in potash and phosphate content where much nitrogen was used.

The point at issue for the commercial man is that the grower using cheap inorganic manures would have a large crop to handle at a low price or, in times of abundance, perhaps no sale at all and, in addition, of low keeping quality. While the grower supplying the soil with organic matter has a moderate crop of high-quality fruit—reducing his handling costs and increasing his returns and, moreover, a crop that will store. In addition, the inorganically grown crop will require more expense in protection against diseases. It seems that the grower spending the most on his manurial treatment will find it the most profitable in the long run.

Another aspect of quality is cleanliness of produce. Blemishes caused by pests and diseases have already been mentioned, but there is the question of dirty produce. Much time and trouble is expended now on the washing of vegetables, with its inevitable abrasion to the skins and loss of keeping quality. Under shallow organic cultivation a type of soil is formed which, although friable, does not splash easily —the crumbs are lightly held together by fungal mycelium and surface algae and no doubt other agencies. Furthermore, the soil does not adhere tenaciously to root vegetables—the soil blocks, being unbroken by cultivation, remain as entities rather than stick to the smooth skin of the vegetable. Washing becomes unnecessary except perhaps immediately after rain, when the earth is soft and may adhere to the fibrous roots. A rinse with the hose is all that is required, without any agitation.

Quality is a much more sensitive criterion of good growing than size of crop. A farmer may use artificial manures solely for the latter purpose, the quality does not matter so much to him—most of his products are processed before reaching the public. Not so for the market gardener. He needs a good crop and it must look good.

A point of interest to market gardeners is that of earliness of crops. It is contended that one must use liberal quantities of quickly soluble nitrogenous manures for this purpose. There is some truth in this but it depends how early. Earliness is becoming something of a snare. There will always be a limited market for out-of-season produce, but with the continuous decrease in purchasing power of the housewife in recent years it is becoming more limited. She is getting rather more choosy. She needs good-quality produce at a reasonable price. This market is unlimited, even in times of glut good quality sells. Unfortunately, many market gardeners and farmers have not seen good quality for so long that they no longer know what it is. When a buyer proves anxious for a grower's produce, then he will know that he has quality. It is a pleasant experience.

A few words on orchard management are appropriate in this chapter.

It is fairly well established now that an orchard can be made self-supporting in organic matter by grassing down. Whether it can be self-supporting in nutrients is a contentious subject, but it is the experience of those practising organic principles that it can be done. A 10-ton crop of apples, for instance, would only remove about 30 lb. of potash and about 6 lb. each of phosphate and lime per acre[1]—not much more potash than a good soil would lose by natural erosion, and apples are generally grown in soils rich in this mineral. So there does not seem to be much point in adding potash.

The third factor required for orchard growth is water. This, in most cases, especially during the months of June, July and August, is the limiting factor, and it would seem profitable to irrigate where possible.

On the other hand, trees are deep rooted and, given a deep soil and a minimum of an inch of rain for each of these months, it should be possible to manage without.

For provision of nitrogen and organic matter it is usual to sow the orchard to grass and clover when the trees are between five and seven years old, though it can be done sooner in wet districts. However, if the trees are mulched with the grass mowings the orchard may be grassed down the same season as planting. Wild white clover and ryegrass or timothy are usual mixtures and, though valuable, it is likely that more deep-rooted species might well be preferable, such as cocksfoot and lucerne. Obviously, if there is a risk of drought, there is a difficulty here, as the deep-rooted crops will compete in the same root zone as the trees. The main object is to provide the ground with as much organic matter as possible without too much competition, and if water is insufficient, deep-rooted plants must be ruled out.

Unfortunately, present-day management of these clover swards does not make full use of their potentialities. The nitrogen aspect was dealt with in a previous chapter (page 115). The futility of repeated cutting to build up organic matter by adding nitrogenous organic matter to the soil in frequent small doses has also been discussed (page 74). In view of Rothamsted's calculation that a tall crop would not transpire any more than short grass, there seems to

---

[1] Analysis of an orchard soil in Herefordshire taken many years ago by Dr. J. A. Voelcker showed 0·72 per cent potash. Assuming a root run of 5 ft. the trees have access to 70 tons of potash per acre, sufficient for 6,000 years.

be no good reason why orchard grass should be mowed at all—except perhaps to encourage white clover where grown, and so long as it does not impede orchard traffic. Mowing only increases the water available to the trees in so far as it restricts the root range of the turf. Shallow-rooted grasses seem to be the economic answer. There might be some immobilization of nutrients for the first season, but thereafter the accumulating organic matter, particularly if leguminous plants were included, would release all the nutrients required. The orchard would not then be existing on the bare minimum of soil capital with its attendant dangers—apart from the saving in expense.

Sir Albert Howard claimed that fruit trees would not make satisfactory growth in India when grass covered the roots, and attributed it to lack of air. It seems more likely that the effect was due to nitrogen starvation, particularly in the heavily leached soils low in organic matter so characteristic of the Tropics. Certainly nitrogen starvation is the governing factor in temperate climates—apart from water. There should be no difficulty if a proportion of legumes is added to the sward. It is, however, true that grassland has a higher carbon-dioxide content than arable—but so has any other land with plant cover. Some concentration of carbonic acid is valuable in releasing minerals, especially on chalky soils.

If competition for water is likely to be severe with a cover crop the difficulty can be got over at the expense of land. Either by reserving a portion of the farm solely for the production of organic matter for use as a mulch around individual trees, say 10 ft. square, or by spacing the trees much more widely apart and growing strips of mulching material down the lanes to be used under adjacent trees. The latter would seem to be the most convenient from the labour aspect—though it would increase labour in other directions somewhat owing to the greater distances involved in other orchard operations. The principle to bear in mind is that water loss in summer under a mulch is negligible, that under clean cultivation moderate, while under grass, whether mown or not, is considerable, especially if deep rooted and long dry periods occur. This, however, is partly counter-balanced by the fact that the water-storage capacity of soil is appreciably greater under turf than cultivation, especially at lower levels, owing to its better structure.[1] Mulching, incidentally, may be of value to other market-garden crops, but is not economically practicable unless the plants have a high cash value. There is a danger that it increases a plant's susceptibility to spring frosts by preventing

[1] W. A. Albrecht, U.S. Dept. Agr. *Yearbook of Agr.*, 1938, 347–60.

warmth from the soil heating the surrounding air by night. Tender plants should not, therefore, be mulched until June.

As regards bush fruit, these are mainly shallow rooted and are plants adapted to semi-woodland and moist conditions. It is not wise, therefore, to allow any competition for nutrients or moisture by allowing grasses to grow right up to the bushes, especially in the case of currants and gooseberries. Raspberries will tolerate a fair amount of growth at the end of the summer, when rainfall is usually adequate. The best method to adopt with these fruits is the same as with tree fruits under dry conditions, a system of growing lucerne, sweet clover, or other bulky material, as a mulch. Either on a separate plot or in widely-spaced rows. One can, of course, purchase straw, sawdust or any other carbonaceous waste for this purpose. If the soil is low in nitrogen, however, there is a risk of nitrogen starvation when sawdust is used, and a light dressing of dung or other nitrogenous material must be put down first. Prunings, with their lignin content, are also useful providers of humus, and should be put through a shredder, if necessary, rather than burnt.

A matter of some influence on soil temperature and moisture is its exposure to wind. A strong dry wind, or even a moderate though cool wind, if the sun is strong, will increase the rate of evaporation from the soil, thus cooling it. This is quite apart from the wind's own cooling effect by day through rapid exchange of the surface layers of air. This latter effect is particularly noticeable for glasshouses where the loss of heat is a direct function of wind speed.

Although the prevailing winds in Britain are south-westerly, the most cooling winds are those from the opposite half of the compass, between north-west and south-east, on account of their generally dry and therefore evaporative nature. This is most true of winds between north-west and north in summer, which are cold and dry. In semi-arid areas wind may also lead to the blowing of soil. Some shelter, then, would seem to be of value where winds are strong or cold.

Shelter is, of course, of value to animals at all times, but more especially in winter.

The best type of shelter belt is one moderately dense above and open below. It has been found that very dense shelter belts cause strong eddies on the leeward side and a rapid build up of air movement.

Figures of wind speed reduction by the better type of shelter belt, about 15 ft. high, is shown below:

Wind speed: 20 m.p.h. at 10 ft. above ground.[1]

[1] L. P. Smith, *Sci. Hort.*, 1950–1, 10, 122.

Position *Relative to Hedge*	Wind Speed *10 ft. Above Ground*
150 ft. up wind	20 m.p.h.
75 ,, ,,	18 ,,
30 ,, ,,	14 ,,
15 ,, ,,	12 ,,
30 ft. down wind	6 ,,
75 ,, ,,	6 ,,
150 ,, ,,	10 ,,
300 ,, ,,	14 ,,
450 ,, ,,	16 ,,
800 ,, ,,	20 ,,

It will be noted that the wind speed is cut to about half at a distance equal to ten times the height of the windbreak. Also that there is a considerable obstructive effect on the windward side.

The best type of tree to use for this purpose is a tall columnar variety such as poplar or Lawson cypress, though actual choice must be modified by the local conditions. This avoids too much shading. The tendency of trees to spread their roots into surrounding land can be checked by biennial treatment with a subsoiler, soil permitting, at a distance from the trunk equivalent to about a third or quarter of their height, according to the anchorage provided by the soil.

VIII

COMPOSTING

The purpose of composting plant and animal residues is to provide the plant with food in a more easily available form than that contained in the original residues. There are, in addition, a number of other benefits to be gained from this preparation of the food.

(a) The food is presented to the plant in a steadily available form, being broken down by soil organisms during the course of the season as and when the plant requires it.

(b) The organic matter provides a useful soil conditioner in a form less readily destroyed than dung, as it contains a better balance of carbon and nitrogen compounds.

(c) Weed seeds and pathogenic organisms are destroyed by the heat and microbial activity of the composting process.

(d) The bulk of the materials used is reduced to about a third, and is in a friable condition, thus considerably reducing handling costs over the original bulk of materials.

(e) The highly concentrated and sometimes harmful compounds in some animal manures are converted to more suitable and better balanced materials for plant use.

(f) The conservation of nitrogen and other easily lost nutrients.

In order that this decomposition of residues shall come about rapidly and efficiently, three factors must be present—namely, air, moisture and sufficient nitrogenous matter, phosphates and trace elements, for the micro-organisms to build up their bodies.

Air, naturally, is required for the micro-organisms' respiration, during which they burn up the carbohydrates in the plant residues with the evolution of considerable heat. If we are to obtain this heat (140–160° F.), the air supply must be adequate. On the other hand, too much air will cool and dry the heap—the supply must be controlled. This is best achieved by wrapping small heaps with some

172

porous material—sacking or slatted boards—or, in the case of large heaps, by situating them out of winds, or, better still, in pits with a network of air channels on the bottom to provide a steady percolation of air from the bottom upwards. It is for the reason of excessive air supply that heaps of less than half a cubic yard are difficult to heat properly except in warm weather, or unless extremely well insulated.

The actual rate of passage of air can vary within quite wide limits; an increase in flow, though carrying away the heat, also gives the organisms the opportunity to work faster and produce more heat, so that the process is to some extent compensatory. Care must be taken to avoid extremes.

In practice, a mixture of soft leafy or grassy material, and plant stems in equal proportions, or straw, with the addition of the commonly recommended one-quarter by bulk of farmyard manure, will give good results. Sawdust, coffee grounds and similar close-textured materials are almost impossible to decompose quickly, mainly because of poor aeration. The addition of sufficient nitrogenous matter and moisture for complete decomposition would effectively stop air from entering these materials. It is better, therefore, if quick results are required, to mix these materials with more fibrous matter.

The loss of nitrogen through composting close-textured materials, particularly if high in nitrogen, is well borne out in the following table.

CHANGES OF NITROGEN CONTENT IN COMPOST HEAPS[1]

Material	Nitrogen at Start lb.	Nitrogen at End lb.	Gain or Loss lb.	Percentage Gain or Loss
Weeds	44·2	25·7	−18·5	−41·8
½ Weeds and ½ Crotalaria	42·8	28·4	−14·5	−33·8
Ditto	49·7	29·2	−20·5	−41·3
Mixed Crop Residues	28·3	29·5	+ 1·3	+ 4·4

Crotalaria is a leguminous plant. Heaps of approximately 2 tons at start.

[1] A. Howard and Y. Wad, *Waste Products of Agriculture.*

It will be noted that the higher the proportion of nitrogen the greater the loss. While a well-aerated heap low in nitrogen finally contained the most. It has been found also that mixed wastes decompose more quickly than single materials partly, no doubt, owing to the improved texture of the heap. But it seems that many micro-organisms thrive better on a mixed diet, and these will be better catered for.[1]

Moisture is necessary and needs to be present within comparatively narrow limits—sufficient to provide a film of moisture for the micro-organisms to travel in, but not so much as to impede aeration or even to run through the heap and cool it. While water passing through will probably not carry away much nutrient, providing the heap is warm and well aerated, should the heap become cold and the micro-organisms inactivated, there is a decided risk of this occurring. In practice, a water content of 50–70 per cent is found to be satisfactory; in effect, the materials will be holding just as much water as they can without any surplus to drain away. Too much water will also impede the activity of fungi which is necessary to decompose the ligneous fractions.

Nitrogen is the chief food element, apart from carbohydrate, required by the micro-organisms. In order to provide a balanced diet for micro-organisms the nitrogen must be between 1·2–3 and 1·8 per cent of the carbohydrate on a dry-weight basis, which represents a carbon-nitrogen ratio of about 33:1. On this basis, a ton of fresh dung containing 10 lb. of nitrogen would suffice for 1,000 lb. of straw already containing about 0·5 per cent of nitrogen. A smaller quantity will result in a somewhat slower decomposition, while a greater quantity will not be utilized by the microbes and will be destroyed by denitrifying organisms or lost to the atmosphere as ammonia or evil-smelling compounds. From the point of view of economy, therefore, it is probably wiser to err on the lower side.

It is interesting to note in this connection that a compost heap is to a large extent self-compensatory in its nitrogen requirements, as micro-organisms feeding on the material can fix considerable amounts from the atmosphere, as shown by the following table:[2]

---

[1] For example: J. T. Auten, *J. Forestry*, 1940, 38, 229; F. G. Gustafson, *Plant Physiol.*, 1943, 18, 704.

[2] A. Howard and Y. Wad, *Waste Products of Agriculture*, 1931, p. 86.

Amount of Dung Used	Total N. at Start lb.	Total N. at End lb.	Percentage Gain of N.	Gain N. in lb.
Full quantity	32·70	34·87	6·6	2·17
¼ quantity	29·12	32·36	11·1	3·24

It will be noted that the low-nitrogen heap has gained 1·7 lb. more of nitrogen over the full-nitrogen heap. This effect is also dependent on the aeration of the heap, as shown below:[1]

	Pit 4 ft. Deep	Pit 2 ft. Deep
Total at start (lb.)	31·25	29·12
Total N. at end (lb.)	29·49	32·36
Loss or gain N. (lb)	− 1·76	+ 3·24
as percentage	− 6·1	+11·1

Both these tables refer to pits containing about 2 tons of wastes. For those farmers having abundant supplies of dung, providing the proportion of this material does not exceed a quarter by bulk to strawy or other low-nitrogen wastes, there will be little loss or gain. For the market gardener, however, who has to buy his dung, only one-sixteenth part of farmyard manure is needed, provided it is fresh, or less if the wastes contain a fair proportion of discarded vegetables, tomato haulms, etc., which are generally a good deal higher in nitrogen than straw. In practice, all that is required is sufficient to ensure satisfactory heating of the heap, permitting a large degree of latitude in proportions of materials used. On this basis, very much less of other nitrogenous materials used for composting than has hitherto been recommended seems necessary. I have myself used such compost for several years with satisfactory results, without, until recently, appreciating the reason for them.

Temperature has some influence on the nitrogen and organic matter content. Less organic matter and nitrogen is lost as the temperature rises from 113° F. (45° C.) to 167° F .(75° C.),[2] and the same applies to temperatures below 86° F. (30° C.). In the intermediate

<hr>

[1] Ibid., 92.

[2] S. A. Waksman, T. C. Corden and N. Hulpoi, *Soil Sci.*, 1939, 47, 83.

range organic matter decomposes more quickly, while the nitrogen content may increase.[1]

The energy food of the organisms, almost entirely carbohydrate, with a little protein, can take many forms varying in digestibility. The most easily decomposed are the sugars which provide food for some nitrogen fixing organisms, notably azotobacter. There are further small quantities of starch and pectins, but the great bulk, as a rule, consists of celluloses and hemi-celluloses which provide food for a great variety of bacteria and fungi. The last group of energy-providing materials are the lignins which, so far as we know, can only be broken down by fungi. Curiously enough, the heat-loving bacteria that live in the heap in its early stages seem to prefer proteins. This possibly accounts for the fact that material low in nitrogenous compounds and activated with sulphate of ammonia or other artificial nitrogen may fail to heat up satisfactorily.

The course of events taking place in a compost heap are roughly as follows. The hemi-celluloses surrounding the cellulose fibres are strongly attacked by fungi and bacteria with the evolution of much heat. This releases the celluloses for attack, mainly by bacteria. This stage is needed after about a week, when the temperature reaches its highest peak. With a gradual fall in temperature the fungi reinvade the heap and its celluloses, and also attack the ligneous fractions, but their action is slow and lasts over several months, according to temperature and aeration. If the material contains as much as 20–30 per cent of lignins, as in wood, decomposition is very slow as the lignin encloses much of the cellulose material, thus protecting it from decomposition. Sawdust, from its finely divided nature, is a good deal more accessible, though this is counterbalanced to some extent by its close texture impeding aeration. The processes are exemplified in the table.

RATE OF DECOMPOSITION OF RYE STRAW[2] AT 35° C.

Duration of Decomposition	0 days	4 days	8 days	24 days	84 days	
Total Organic Matter	100	97	78	62	49	As percentage
Celluloses	57	56·1	41·9	30·4	24·2	of original
Hemi-celluloses	21·85	18·35	10·81	8·42	6·85	organic matter

[1] A. G. Norman, *Ann. Appl. Biol.*, 1931, 18, 244.
[2] Figures extracted from A. G. Norman, *Biochem. Journ.*, 1929, 23, 1367.

Plate 21

MULCHING UNDER WARM DRY CONDITIONS
*

(*a*) Runner beans mulched with grass and clover from the headlands.
(*b*) Tomato plants under cloches.
4 Plants on the right mulched.
4 Plants on the left not mulched.

Plate 22

SOFT FRUITS UNDER ORGANIC METHODS
*

(*a*) Raspberries (Var. Malling G) grown in grass, cut before picking and used for a mulch. Measuring rod is seven feet high.

(*b*) Raspberry and blackcurrant plantations under lucerne. Four year old plum on the right. Straw mulch previous three years. Soil analysis—phosphate low, potash very low. Measuring rod seven feet. Some canes are over eight feet high.

The use of earth in the compost heap, while not essential to the decomposition, has beneficial effects on the type of decomposition and the absorption of nitrogen. The ideal soil to add to the heap is a calcareous clay to keep the reaction neutral. During the course of decomposition a number of organic acids are produced, which, though eventually oxidized to carbon dioxide may impede the activities of fungi in decomposing lignin and also the cellulose-decomposing organisms, resulting in a rather coarse compost. For some types of soil, particularly the heavier ones, this type of compost may be preferable to a fine-grained one, as it will assist in keeping the topsoil open. Ground limestone may be used for this purpose, but soil contains other minerals besides calcium which are useful in maintaining the neutrality of the heap and are also plant nutrients. Though these may be in a form not readily available to the plant, the intense bacterial activity and competition has every opportunity of rendering them so. Clay and organic matter can combine together to form complexes which have been considered to be of value, but experimental work has not shown them to have any special effect on soil structure.[1]

The clay fraction of the soil performs a useful function by absorbing any ammonia evolved,[2] and also by coating the coarser materials with a moisture-holding layer which aids the activity of micro-organisms. For the latter reason, and also to facilitate mixing, the earth is best applied as a thin slurry. Only small amounts are needed, just enough to colour each layer as the heap is built up.

The construction of a compost heap is commenced on an earth base 6–10 ft. wide and as long as convenient. The narrower width may be better where the work has to be done from one side of the heap. If the heap is to exceed 2 ft. in depth some system of air channels under the heap will have to be provided so that no part of the heap is more than 2 ft. from an air supply—the 3-in. diameter drainpipe is a convenient unit.

A layer of coarse material, hedge trimmings or similar material, is laid down first to permit air to make its exit from the pipes, and a layer of vegetable wastes 6–9 in. thick, according to its density, is laid on top of this. This is spread in turn with a 2–3-in. layer of dung and sufficient water added, if necessary, to saturate it. With this dung a small quantity of earth or ground limestone may be spread, and a small amount of old compost is advisable also, to ensure that the

---

[1] A. P. Mazurak, Soil Sci. Soc. Amer. Proc., 1949, 14, 28–34.
[2] C. N. Acharaya, *Indian J. Agric. Soc.*, 1945, 15, 214; 1946, 16, 90.

heap contains ample micro-organisms of the right type for a rapid build up.

A further 6–9 in. of strawy wastes is laid on top and spread in turn with dung, earth and old compost. The heap may be built to about 6 ft. high—or higher, if light bulky materials are used.

While it is not necessary to mix the dung and earth to a slurry with water, it will be found that much more intimate mixing of the materials is achieved, with a consequent saving in the amount of turning required later. This is particularly true where the amount of dung available is limited.

As an approximate guide to quantities, a satisfactory product is obtained by mixing about 30 lb. of fresh dung and 5 lb. of earth with a pound or two of old compost, in water, in a 20-gallon tub, and spreading the mixture over every layer of a heap 10 ft. by 6 ft. consisting mainly of roadside weeds and hedge trimmings. These proportions are open to wide variations and there is scope for a good deal of experiment as to the most economic methods and proportions under local conditions.

Some farmers save the urine from cattle sheds and milking parlours, which is almost certainly best utilized in the making of compost, particularly where dry straw is used. Others, who keep cattle in covered yards, may predigest their compost by spreading straw and other farm wastes in the yard with a little earth or limestone, where the trampling of the cattle breaks up the fibres of coarse materials (kale stems, etc.) and efficiently mixes the dung with the wastes. This mixture is periodically removed and stacked, when strong heating develops. The use of open yards is not entirely satisfactory where there is much rain, losses of nitrogen and other nutrients are very heavy.

The close-textured forms of dung such as pig manure need a plentiful admixture of bedding or stalky materials to aerate it and when so treated are as easy to handle as any. Poultry in deep litter provide yet another means of mixing compost materials, without labour, and a ton of poultry manure, containing 40 lb. of nitrogen, will need at least 2 tons of straw to avoid loss of this element.

Sir Albert Howard and Y. Wad, in their book *The Waste Products of Agriculture*, give very precise quantities of materials for use under Indian conditions, including quantities of urine-soaked earth from cattle sheds and details of the exact timing and quantities of water to be applied. Woody materials were crushed by the farm traffic in the roadways. Useful details of compost making under present-day circumstances may be obtained from *Compost for 1,000 Acre Farm or Garden Plot*, by F. H. Billington (Faber & Faber).

It is perhaps unfortunate for the ordinary farmer that economics are a prime consideration, particularly where labour and machinery are concerned. His criterion is inevitably what must be done, rather than what is desirable, even at the risk of some loss of quality. It is fortunate, therefore, that there is such a wide degree of latitude in quantities and methods, so that, providing a few simple conditions obtain, quite wide variations in proportions will provide valuable material. Indeed, as all organic matter eventually decomposes to humus compounds having a C.-N. ratio of 10 : 1 under aerated conditions, we have little control over the final composition of the compost—we can only control its rate and, to some extent, the degree of decomposition.

It is on the question of labour that the matter of turning of the heap largely rests. It is desirable after two to three weeks to turn the heap to moisten, mix and aerate it and place all the unheated outside portions to the centre. This is best done by cutting the heap down into vertical sections and rebuilding on an adjacent site. A further turn after another two to three weeks may be followed by a third turn at two months. It seems likely that, providing the heap can be well mixed at the start, kept moist and aerated, the need for turning largely disappears.

The undigested portions on the top and sides of the heap can be cut away and provide a useful activator for the next heap, or, if straw bales are available to surround the heap, they will serve the double purpose of ensuring complete decomposition of the heap and excellent insulation. The remains of the bales will provide ideal material for the next heap.

Some mechanical means to chop up the coarser materials and mix the dung and plant residues would be an improvement, but even the application of the dung as a slurry gives a very good mix.

To maintain aeration in large heaps vertical holes may be driven into the heap with a crowbar 3 to 4 ft. apart, or posts used during building and removed on completion. A fairly high proportion of coarse materials is also required.

I have myself frequently used unturned compost, and though no comparative figures are available there was no noticeable difference in cropping.

The moisture content requires little attention in Britain at least.

Initially, dry material, such as straw, however, requires about 200 gallons per ton to soak it, and a further 400 gallons by instalments of 100 gallons at about fortnightly intervals. Most composting is done during autumn and winter when excessive moisture is the rule. A heap older than three months should be protected from rain, in any case, and from heavy rain at all times.

Under warm conditions, as in the temperate summer, compost is ready for use in three months, when it becomes crumbly. It is, unfortunately, a wasting asset and slowly loses nitrogen and organic matter after this time. The only way to preserve it is by drying, which inhibits bacterial activity.

LOSSES OF NITROGEN IN COMPOST HEAPS AFTER THREE MONTHS[1]

(Nitrogen content per cent.)

Heap	*Three Months*	*Four Months*
7	0·90	0·88
8	1·00	0·93
14	0·84	0·81
15	0·72	0·68

In winter weather the process will take longer, perhaps six months. The initial stage of intense heating is unaffected by temperature conditions but is affected by heavy rain and very much by wind. Large heaps exposed to a winter storm go cold in a night and often fail to reheat. Hence the need for protection in pits or some sort of screen. A very efficient protection is provided by baled straw, which becomes partly rotted itself and a valuable addition to the succeeding heap.

Much compost has been made, particularly at our research establishments, by soaking straw in solutions of nitrogenous salts such as sulphate of ammonia, nitrate of potash, calcium cynamide, 'Adco' and nitro-chalk. The greatest difficulty with such heaps is to keep them moist. Even in my own humid area, with considerable rainfall, rotting has been consistently uneven with such materials, giving patches of dry unrotted material at the centre and lower layers of the heap. A good deal of extra labour and attention is needed to ensure that they are properly moistened. The reason for this difficulty is probably the absence of colloidal substances which are so abundant in dung and hold moisture on the smooth, hard surfaces of strawy materials. The use of clay slurry in such heaps would be useful. Such

[1] A. Howard and Y. Wad, *Waste Products of Agriculture*, 1931.

composts are normally considerably inferior to those made with dung, possibly because they have an initially higher content of freely available nitrogen, which is consequently liable to rapid losses,[1] a matter which has been explained before. I have found this to be so in comparative trials with potatoes, the compost being about equivalent to a similar quantity of dung (page 59).

A properly made compost of farm wastes usually contains nutrients in the following range—potash 0·8–1·0 per cent, phosphate 0·4–0·6 per cent, and nitrogen 2·0–4·0 per cent. Wide variations of minerals can, however, be expected according to the nature of the constituents. That made from sewage and household waste has a similar range of nitrogen and phosphate but is generally lower in potash, as much of this valuable constituent is lost in the liquid portion of the sewage. It has also a much higher content of mineral matter—cinders, etc.—and, consequently, a relatively lower content of organic matter. It may, therefore, be of particular benefit for clay soils.

A few costings are available for the manufacture of compost; probably the most comprehensive is that given by Mr. Friend Sykes in *Humus and the Farmer*. Making and spreading compost in large quantities with the aid of a crane and grab, manure spreader, water pump and haulage machinery, and including depreciation, labour, materials, etc., and spreading, works out at 4s. 6d. a ton, at 1945 prices. Ten shillings would be nearer to-day's cost.

On the other hand, compost in a market garden buying all its materials—straw, dung, hoof and horn, and lime—and using hand labour, worked out at 30s. a ton exclusive of spreading. In my own case, collecting free roadside wastes, would work out at around 15s. a ton, to which must be added, say, 5s.-worth of dung, and a further 10s. a ton for mixing by hand. Carting and spreading, again by hand, would cost a further 10s.—a total of 35s.

The farmer normally has straw to dispose of, and its cost is simply that of handling, but the market gardener usually has to obtain his materials from outside sources. Their variety is great; sawdust, paper, spent hops, coffee grounds; and, in fact, any material containing a large percentage of carbohydrate is suitable providing it contains no antiseptics or other substances inimical to micro-organisms; for example oil or creosote, and compounds of sulphur or copper. The main factor is their cost of transport, so the supply is largely limited to locally-produced materials.

A growing interest is being taken at present in the utilization of

---

[1] A. Howard and Y. Wad, *Waste Products of Agriculture*, 1931, p. 87.

town refuse and sewage as a compost.[1] The salvaging of the immense amount of plant nutrients and organic matter wasted in the destruction of these materials is a matter of some urgency.

The recent statement by Prof. Stoughton of Reading University, at the Horticultural Congress of September 1952, gives food for thought. He is working on the subject of composting town waste and gives the following figures. The quantity of household refuse produced in England and Wales in 1937 was 8,000,000 tons, costing £5½ million to dispose of, while the London County Council, at its northern outfall alone, had some 2,000,000 tons of sludge to dispose of, mostly carted away in ships and dumped in the North Sea. The cost these days must be two or three times that of 1937, and the cost of installing and maintaining all our sewage-disposal plants must be astronomical.

According to more recent figures by Mr. J. C. Wylie,[2] there is a potential output of between five and seven million tons of compost from sewage, with a nitrogen content of about 42,000 tons (0·6 per cent). This represents a quarter of the synthetic nitrogen used on the land in this country (of which two-thirds are lost in the soil), so that with reasonably conservative methods of agriculture, we could be self-supporting in this element through the use of compost alone. The phosphate level is around 0·25 per cent, and potash 0·17 per cent, for the moist product. A fair amount of mineral matter (cinders, earth, etc.) is inevitable in these composts, ranging around 50 per cent. While these are not harmful, they add to the transport costs, so that for economic reasons a number of small composting units throughout the country would seem more economic than a few large centres.

There may in some cases be harmful impurities in the composting materials which remain unaltered. Most sewage composts seem to contain rather high levels of zinc and lead, especially the former. There are also large quantities of detergents used these days, of which the majority seem harmless though the cationic types used in disinfectants are harmful to bacterial action. While the use of composts containing these materials, and perhaps other industrial wastes, would not be likely to cause harm on open land, there is a risk of a build up in a glasshouse soil.

[1] *Interim Report on Composting* and *Fertility from Town Wastes*. Not yet published. Provisional Title (Faber), J. C. Wylie, Engineer to Dumfries County Council. Also *How can we use our Sewage and Refuse?*, Albert Howard Foundation.
[2] J. C. Wylie, *J. Roy. Soc. Arts*, 1952, Vol. c, 4874, 499.

Composting

The process of composting household refuse and sewage follows the broad principles of composting farm wastes.[1] The sewage is allowed to settle and the resulting sludge high in nitrogen is pumped on to the household refuse which has been sorted over for salvageable materials and crushed, to maintain a carbon-nitrogen ratio of 30:1. This mixture is allowed to drain and turned to admit air. It is generally found necessary also to add coarse strawy wastes or roadside trimmings to assist aeration, and some limestone to check acid conditions. Rapid and intense heating develops, and after six weeks the compost is turned out to ripen, which takes a further six weeks.

A number of municipal authorities are experimenting in this direction, notably Dumfries[2] and Edinburgh, in Britain; while several others have been in progress since before the war, including the West Middlesex Authority and Leatherhead. Even if the sale of the compost has not always shown much profit on the installation of composting equipment, at least it is not a complete dead loss to the community, either financially or in fertility.

In Holland a government-supported scheme costing about a million pounds is being started to build composting plants in various large towns. One large concern, at Wyster, already produces 120,000 tons of compost annually, at 5s. 6d. a ton, and is run by fifty-five workers. The town inhabitants contribute about 2s. 3d. per head annually towards the upkeep of the works. Not an excessive charge for disposal of sewage and garbage.

Figures given by Mr. J. C. Wylie,[3] County Engineer for Dumfries, show that the compost in their small plant sold at 50s. per ton in bulk or 10s. per cwt. dried and bagged.

In overseas countries, especially S. Africa and India, municipal composting produces hundreds of thousands of tons annually.

It is generally considered that compost should be applied to the soil when it has reached a crumbly consistency. It has then lost most of its energy-rich compounds with the exception of some of the lignin fraction, so that the bacterial protoplasm will, to a large extent, die off, releasing plant foods from their bodies for immediate use of the plant. The humic material that has been formed is extremely valuable for conditioning the soil; that is, rendering it crumbly as against the two extremes of hard clods or sandy dust.

It seems reasonable to assume that, under some conditions, com-

---

[1] J. C. Wylie, *Composting*. Reprint from *Public Cleansing and Salvage*.
[2] *Kirkconnel Refuse and Sewage Treatment Plant*, Dumfries County Council.
[3] *Interim Report on Composting*, Dumfries County Council.

post could with benefit be applied to the land as soon as it is reasonably fit to work with, and has been heated to kill weeds and parasitic organisms. It would still contain much energy-rich material which, by rotting in the soil, would provide a great deal more food for soil organisms than the completely rotted product. In this way, a much greater increase of soil micro-organisms would be achieved, and, consequently, a greater release of minerals and a larger worm population.

Most worms, as shown before, can only digest carbohydrates as sugars and starch in any quantity—and these are absent from compost. They prefer fresh material. There are, of course, one or two specialized worms which live in dung and compost by feeding on bacteria and fungi rather than the compost itself. These species die when the compost is applied to the soil and their normal food becomes exhausted.

It has been shown that readily decomposable organic matter brings about a rapid improvement in soil-crumb structure. This effect is not obtained by the use of well-rotted composts or farmyard manure[1] except as a relatively long-term cumulative effect of organic matter.[2] This is due to the immediate production of a high-fungal population. The effect is, however, temporary.

It is for these reasons that sheet composting is a far more economical method whenever practicable, quite apart from the immense saving in labour. In·support of this view, in an experiment using sewage and straw composted versus sewage and straw ploughed in,[3] it was shown that spinach beet was just as good sown directly on this mixture as on the compost. In both cases, nine weeks were allowed for decomposition. It was found, furthermore, that the compost had lost 23 per cent of its organic matter and 12 per cent of its nitrogen. The greater the amount of decaying, rather than decayed, material in the soil the greater the total release of plant food and the longer its duration.

It should be pointed out, however, that there is a danger here. If immature compost low in nitrogen is ploughed in, some nitrogen starvation of the crop is almost inevitable, as all available nitrogen will be mobilized by the micro-organisms feeding on the compost. But if left on or near the surface it will be taken down little by little by the worms and surface fauna. Perhaps slight temporary and local immobilization of nitrogen will occur as each piece of organic matter

<hr>

[1] J. P. Martin, Proc. Soil Sci., Amer., 1943, 7, 218.
[2] Rothamsted Expt. Sta. Rep., 1947, 33.
[3] C. Bould, *Sci. Hort.*, 1949, 9, 23.

is drawn into the ground, but the phase will soon pass, freeing the contained nutrients. There will be a rate of release of nutrients more rapid than that at which they are temporarily immobilized, giving a net effect of gradual release.

Whether bio-dynamic herbal activators are of real value in the compost heap is hard to say. Those who use them claim they are. My own experience is limited, but there did not appear to be any superiority over compost without the activator.

It may well be, however, in some cases, that some herbal preparations will provide a useful material not already present in the compost heap, but in the usual mixture of farm and garden wastes this seems unlikely. My own compost heaps, in any case, contain considerable amounts of plants commonly used in these preparations.

Certainly the bio-dynamic school have achieved interesting results in many fields and they are deserving of further investigation. Such results may be explained on catalytic or enzymic grounds.

There does not seem to be any risk of spreading disease when composting diseased material. The heat evolved is sufficient to destroy any living or resting stages of pathogenic organisms, especially when one considers that this temperature of 140–160° F. is maintained for for two to three weeks. Generally speaking, any active organisms will soon die on the decomposition of their hosts, and, in any case, other predatory organisms in the heap will make short work of them.

An interesting experiment on composting and virus was made by the John Innes Horticultural Institute. Details are given in a little book published by them entitled *Answers to Growers*.

A quantity of the most virus-infected plants on the place were pulled and allowed to rot with the addition of a little lime. A few plants were grown in pots on this 'compost'. Growth was unexpectedly good and no virus was seen. Some virus infection was found in the roots only. This sounds odd, but root infection does not always travel to the tops in tomatoes.

A further experiment was made growing tomatoes in the open, giving different batches of plants compost or dung at 20 tons per acre. The plants on compost gave slightly higher yields and were healthier. though the difference was not considered to be significant. At any rate it was shown that compost from diseased tomatoes was quite safe to use for tomatoes.

In Ceylon, investigations on the destruction of hookworm and roundworm eggs in composted sewage were made.[1] No living eggs were ever found.

The late Mr. A. R. Wills of Romsey, of a large concern of farms and market gardens, composted tomato plants badly attacked with verticillum wilt. This was used to grow the next tomato crop with outstanding results.

It may well be that some of the freedom from disease and weeds enjoyed by organic farms over the orthodox is due to the composting of dung before its application to the land.

Organically activated compost is commonly believed to be of twice the value of dung, weight for weight. In a comparison between different sets of applications of dung and compost to potatoes made by myself it was certainly true. The reason is fairly obvious—there is a much better utilization of nutrients from compost than dung. We have already seen the losses of nitrogen that occur from 'straight' dung. Dung is, at best, an ill-balanced fertilizer, containing a large amount of easily available nitrogen, the use of which may lead to soft sappy growth liable to disease. If, on the other hand, it is used well rotted, or allowed to rot in the ground, the greater part of the nitrogen will have been lost to the atmosphere, and in turn the excessive nitrogen will hasten the decay and loss of any other readily available organic matter in the soil. Composting dung, by mixing this excess nitrogen with carbohydrate, ensures its preservation, along with many other of its soluble constituents. The dung-heap can lose over 50 per cent of these constituents in three months.[2]

Compost, on the other hand, releases its nitrogen slowly as the materials of which it consists are broken down. Too slowly, in fact, in the view of some authorities. This was Rothamsted's conclusion in comparisons of composts and dung. But when one considers the small quantities used (see straw ploughed in and composted experiments, page 78), the kind of composts used, and the use of nitrogenous manures with them, this is hardly surprising. They do not yet appreciate the need for building up the organic content of the soil to a high level which necessitates withholding the destructive influences of nitrogenous fertilizers, and the addition, for a period at least, of large quantities of carbohydrate material to the soil. Further, the natural structure of the soil and arrangement of organic matter in it must not be disturbed.

A word on the time of spreading well-rotted compost. As it is a readily available plant food—although perhaps a more slowly avail-

---

[1] Nicholls and Gunawardana, 'The Destruction of Helminth Ova in Nightsoil by Composting', *Ceylon Journal of Science*, Vol. V, Part I, 1939.

[2] R. M. Salter and C. J. Schollenberger, Ohio Agr. Expt. Sta. *Bull.* 605, 1939.

able one than the usual run of fertilizers—it should be spread just before the crop is sown or planted and lightly cultivated in. Unfortunately, the ideal is not always possible. Most composting is done during slack time in autumn and winter and is not ready until spring. This fits quite well with spring-sown crops, the compost can be applied about March, giving it time to incorporate into the soil before sowing. Autumn-sown crops could quite feasibly be given a light dressing at any time during the winter or spring, providing the ground is reasonably dry. There is a risk of smothering a few plants, but not a serious one. Grassland can be similarly treated, though with the deep-rooted, mixed, leguminous leys recommended it should not be necessary to compost them. They are self-sufficient.

There is considered to be a risk that, if compost is applied early in the winter, there will be losses of nutrients by leaching. The problem does not normally arise, but it is well to mention the fact that when soil colloids are destroyed by freezing they will release the nutrients they have absorbed to be washed away by rain. The possibility was brought up by a writer in the *Soil Association Journal* after an article on the subject of the effect of freezing on soil colloids, by Mr. A. F. Wiess, B.Sc., in Vol. 2, No. 1. It seems that this destruction of soil colloids by freezing only occurs in the presence of electrolytes, i.e. soluble salts. If these are not present the colloids recover on thawing. This damage would not occur to organic compost spread on the soil in winter—though it would be much more likely with a compost activated with inorganic salts. There are a number of reasons for this. Plant nutrients are not normally present in the inorganic state in compost, but in the form of proteins, as bacterial and fungal protoplasm. Whether these are destroyed by freezing in some cases is open to speculation; even if they are, their breakdown products are not likely to be simple inorganic salts, and will be reabsorbed by the hardier types with a few hours of milder weather. Even if there were inorganic salts present, any rain prior to the frost would soon remove them. The soil sample used to support the view was from a new lawn and may have contained a fertilizer, but in any case the specimen appears to have been oven-dried to 100° C., at which temperature considerable changes will have occurred, including the breakdown of all living matter and the development of electrolytes under such conditions is very likely. Even under such conditions, mineral salts, with the exception of nitrogen, are quickly recombined with soil minerals.

However, most of us will avoid the risk anyway by spreading

compost in late winter or early spring if only as a matter of convenience.

An interesting point which may as well be discussed here is the fermentation of manure to generate methane gas on the farm for heating and lighting. It is claimed that the manure does not lose anything in the way of plant nutrients by this treatment. In the narrow sense this is true—but its energy food has been lost and it is therefore of little value as a soil food, i.e. for the micro-organisms. It is, in fact, in much the same category as any other nitrogen-rich fertilizer and should only be used on the compost heap or, in moderation, on grassland.

A few words on the principles underlying 'sheet composting' may be discussed here. The object in view is to decompose plant and animal residues in the soil, with the consequent saving in labour, over the making of compost. The danger of incorporating plant residues in soil is that they often require a good deal of nitrogen for their decomposition, and a crop growing in such soil is deprived of nitrogen by the organisms decomposing the residues.

The table illustrates the amount of nitrogen taken from the soil after three months by 100 grammes of various plant roots during their decomposition.[1]

Material	Nitrogen Content Per Cent	Nitrogen Immobilized mg.
Oat Roots	0·45	569
Timothy Roots	0·62	567
Maize Roots	0·79	573
Clover Roots	1·71	63
Dried Blood	10·71	released 1,434

Though the nitrogen content of the first three varies, the amount of nitrogen immobilized is very similar. The reason lies in the relative resistance of those materials to decomposition. Maize, a woody root, though containing more nitrogen, has immobilized nitrogen for a longer period—it is likely that the oat and timothy-grass roots were already fully decomposed after three months and were releasing some of their nitrogen again. It is not quite safe, therefore, to calculate the

[1] Figures calculated from an experiment by S. A. Waksman and F. G. Tenney, *Soil Sci.*, 1927, 24, 317.

amount of nitrogen that will be immobilized from the nitrogen content of the material—immobilization reaches a certain peak which varies in time according to the material. However, it is a fairly reliable guide. Clover roots, on the other hand, immobilized very little nitrogen—their content of 1·7 per cent is very near the level at which no immobilization can occur. Dried blood, as a comparison, contains 10 per cent of nitrogen and released far more than it immobilized.

In order to ensure that the plant does not go short of nitrogen when plant residues are incorporated into the soil their nitrogen content must be raised to around 1·8 per cent. This is conveniently done by adding dung to the pasture or crop remains before cultivating them in. A pasture generally provides something like 1 or 2 tons of dry matter per annum, with a nitrogen content averaging 1 per cent, so that an addition of 20 to 30 lb. of nitrogen per acre should suffice. This represents 2 or 3 tons of dung. This small application could well be applied by a heavy stocking a week or two before breaking up.

Sheet composting is a very profitable means of disposing of the straw left after combine harvesting, and is best carried out as soon after harvesting as possible. The quantity of straw is usually of the order of 50 cwt. per acre, with a nitrogen content of 0·5 per cent. This is fairly easily decomposed and as the crop sown will not require any nitrogen until the following spring it will only be necessary to raise the nitrogen content to about 1·5 per cent. This will require some 50 lb. of nitrogen or 5 tons of fresh farmyard manure per acre, preferably well mixed into the surface soil with a rotary cultivator. A mature mustard crop, kale stems and similar residues, may be treated in the same manner.

While these dressings of farmyard manure seem unusually small, it must be remembered that they are economical dressings—little nitrogen, the chief food element of value in dung—is lost. Whereas with the normal application of 15 tons to arable land some 70 per cent is lost—because there is no carbohydrate for the micro-organisms to live on and absorb the nitrogen. Indeed, there is less risk of nitrogen loss by sheet composting than in the compost heap (see page 114), provided the dung is fresh and it is turned in immediately —otherwise serious loss of nitrogen will occur and larger applications must be made. The effect of these losses on the ensuing crop is shown on the next page.[1]

[1] R. M. Salter, and C. J. Schollenberger,, Ohio Agric. Expt. Sta, Bull, 605, 1939.

	Relative *Yields* *per cent*
Manure spread and ploughed in immediately	100
Manure spread 2 days before ploughing	71
Manure in piles 2 days before spreading and ploughing	80
Manure spread 14 days before ploughing	49
Manure in piles 14 days before spreading and ploughing	55

Manure is often left in heaps for months in our farming system.

If the application of the dung has to be delayed until spring, somewhat larger amounts are advisable, as much of the nitrogen added will be temporarily absorbed by the micro-organisms digesting the straw. An increase of 50 per cent should suffice—again with the foregoing provisions. One of the major characteristics of organic method is their economy. Even so, should a greedy crop, like kale, cabbage or market-garden crop be planted immediately afterwards, faster growth will be obtained by increasing these amounts. Such plants are able to absorb nitrogen quickly—though some loss is inevitable.

IX

DISEASE RESISTANCE IN PLANTS

This chapter must of necessity be largely surmise owing to the small amount of research, and that mostly indirect, which has been done on the subject.

For convenience, plant diseases, as indeed of animals also, may be divided into food-deficiency diseases and those caused by pathogenic organisms. Where one ends and the other begins it is impossible to draw the line. In fact, the distinction is quite artificial—the second is very often a result of the first.

However, it is only intended here to deal with the pathogenic organisms, the pests and diseases.

It is commonly observed amongst farmers and gardeners that the normal healthy-looking plant is not so readily attacked as the thin, pale or stunted specimen, or the soft, lush, fast-growing one. Either extreme indicates unbalanced nutrition in one or more respects.

As it is much easier to feed a balanced diet organically than in-organically (and the latter only balanced according to our lights), it seems that the organic farmer is off to a good start. It will be inter-esting to try to find possible specific reasons for the incidence of certain pests and diseases.

The first pests that ceased to give me trouble on changing to organic methods were the sucking insects: greenflies or aphides and red spider.

Recent research on the greenfly and its habits[1] show that it prefers either very young leaves or old, dying leaves. Those in the middle of the stem it leaves alone. These two extremes have one thing in com-mon —they are richer in nitrogenous matter—protein, relative to carbohydrate—than are the photosynthesizing leaves along the centre of the stem. That the aphis doesn't want all the carbohydrate it

---

[1] J. S. Kennedy, *Aphides and Plant Growth* (New Biology 11.)

sucks up is evident from the quantity of sugary honeydew that it excretes. For the active or warm-blooded animal, like the bee or the bullock, a good deal of carbohydrate is needed. But not so the sedentary greenfly or red spider. Their sole interest is rapid reproduction, which makes a big call on protein. (This may be the reason for the observation that 'compost grown' vegetables are left alone by some pests although more palatable to animals—they contain more sugar.)

In an experiment on the cabbage aphis,[1] it was found that during the months March to October the higher the nitrogen content of the cabbage the faster the aphides reproduced. During winter, when the amount of sugar photosynthesized by the cabbage was low, it seemed that there was often insufficient for the insects' needs. It therefore reproduced fastest when the sugar was comparatively high. Another interesting point was that as the protein content of the plant fell below a certain level winged forms of aphides appeared, ready to find a fresh host.

These aphides also exert some control over their food supply by some means; their activities cause a rapid reduction of carbohydrates in the plant, with a relative rise in the proportion of protein. Once the plant is badly attacked it seems doomed, unless some other agency comes to the rescue.

It seems, then, that a plant having a sap rich in amino-acids or other nitrogenous products will be liable to greenfly, and it is just this soft, dark-green, fast-growing plant fed on rapidly soluble nitrogen that gets the greenfly. It is now common knowledge that nitrogenous manures do increase the protein content of plants. So the danger of their application, more particularly as the quickly soluble inorganic form, becomes apparent.

Here at any rate is a possible explanation. But greenfly also increase in dry weather. Partly because of the very favourable weather conditions, no doubt, and yet in hot dry weather the concentration of sugar increases in the cell sap. A possible explanation would be that by the gradual drying out of the upper layers of the soil a large quantity of micro-organisms would be brought to a standstill and their active stages slowly die out. Their nitrogenous remains would then be absorbed by the plant before the soil dried out completely. Nitrate certainly accumulates in the soil during dry weather. The net effect, anyway, would be the same as a dose of nitrogen to the plant.

Even this is not the whole story. It was frequently observed by Sir

[1] A. C. Evans, *Ann. Appl. Biol.*, 1939, 25, 558–72; 1941, 28, 368–71.

DISEASE RESISTANCE UNDER ORGANIC METHODS

(*a*) Home Saved Seed Potatoes.
Fourth season from Scotch certified seed. Crop of 14 tons per acre after an annual crop of sweet clover. Note the clean skins though grown in a calcareous soil. Variety Majestic.

(*b*) Apple, Egremont Russet, a scab-resistant variety.
Left, from a tree grown under clean cultivation and manured for vegetables with a light dressing of dung and a high potash complete fertilizer, scab resistance broken down, many rotted in store.
Right, tree grown 12 yards away in grass and lucerne, uncut. No scab. Kept well in store.

POTATO EELWORM TRIAL

Left, fresh compost at 30 tons per acre, soil undug.
Right, potato fertilizer 10 cwt per acre, dug. Variety Majestic.

(*a*) 17th June. The compost plants after a slow start have caught up with the fertilizer plants.

(*b*) 26th July. Fertiliser plants dead. Compost plants beginning to die back. Note preponderance of annual weeds on the dug and fertilized plot.

(*c*) 30th August. Crop harvested. Composted 7 lb 4 oz. Fertilized 4 lb 12 oz. Eelworm cyst counts, Mid-July and November.
Compost 22 and 19.   Fertilizer 134 and 26.

Albert Howard, and indeed I have noticed it myself, that plants growing in a poorly-aerated soil are also liable to greenfly. Speedwell —a common weed on my land—when growing under cloches in autumn or spring is frequently infested when it takes root near the cold, wet soil at the edge of the cloche, but plants in the centre grow much more vigorously and remain free. Theoretically, the soil at the edge should be deficient of nitrogen, on account of the very large amount of water that runs through it off the cloche. Perhaps the kind of nitrogen absorbed is different, as ammonia or amines, rather than nitrates, as occurs in well-aerated soil, or some totally different mechanism may be involved. Or perhaps the carbohydrate manufacturing mechanism is disturbed. The moral, it seems, is to improve soil conditions.

This is not to minimize the great part played by the aphis's many enemies. These include ladybirds and their larvæ, hover-fly larvæ, lacewing flies, anthocorid bugs, spiders, birds and mites. They also have a bacterial disease and a fungus which literally roots them to the spot. These insects, if given a free hand, will always get on top of an infestation in a week or two without, as a rule, much damage done. Dr. Massee of East Malling has several times recorded his view that we spray our orchards too much and don't give these creatures a chance.

There was, during 1952, a fairly heavy infestation of greenflies on my blackcurrant bushes during May, which lasted about a fortnight, and some leafcurling resulted. By the end of this time only thousands of empty skins sucked dry could be seen. It is worth noting that for the first time for some years I clean cultivated between the rows prior to sowing lucerne. Perhaps the sudden flush of nitrogen resulting from this action had something to do with the attack.

Some pests normally live on dead and decaying material. Such, for instance, are the wireworms, leather jackets, slugs and snails. There is the story of the inorganic farmer with 2 million wireworms to the acre having a damaged crop, while his organic neighbour had a full crop with 5 million wireworms to the acre. The organic farmer, inadvertently perhaps, fed his wireworms with organic matter and they left his crop alone. It is well known that these pests are worse the second year when the turf has disintegrated.

Slugs, it has been shown, are normally feeders on any dead material, from dead worms to manure. But, again, if this gets short, by clean cultivation, they have to feed on the crop. A number of organic gardeners practise leaving the weeds to decay on the spot, to feed the slugs, and no doubt birds come and clean them up later. When I tried the experiment under cloches in winter, the slugs ate a good deal of the weed which, unfortunately, took root and grew again. The result was a mass migration back to the lettuce. In any case, in this area, a weed takes root again at any time of the year on account of the cool humid climate. It seems better to depend on 'Meta' and ducks.

Slugs appear to prefer wilted material probably because its starch is partly converted to sugar and it tastes sweeter. This would account for its partiality to freshly-planted seedlings. The answer is not to let the seedlings wilt, and the organic gardener scores again. Slugs, incidentally, are found on farmland in quantities of 2 to 3 cwt. per acre.

Caterpillars, though present, do not seem to devastate cabbages to the extent sometimes seen: of plants reduced to skeletons. I have watched white butterflies day after day in August over cabbages and expected to find them full of caterpillars. As a rule, only one or two caterpillars have developed in the large plants and just an occasional one on young plants. The amount of damage has usually been quite small—though often enough to spoil the crop's appearance for sale. This seems to be a general observation. Whether, again, the caterpillar prefers a high-protein cabbage or whether the smell is not so strong if grown organically, we cannot say at present. Even so, it seems advisable to keep a tin of derris on the shelf, as damage is sometimes serious. Having seen oak woods stripped in June—and certainly there was no artificial manure or deep cultivation used there—complete disarmament seems unwise. My apple trees, too, have caterpillars: again, they are not stripped, but a damaged leader is not easy to replace.

Even so, their natural enemies are considerable. Take this account of the destruction of Cabbage White caterpillars.

Of 10,000 caterpillars hatched 59·17 per cent are destroyed by disease; of the 4,083 left 84·21 per cent are destroyed by *Apalantes* parasite; of the 645 left 27 per cent are destroyed by disease as pupæ; of the 471 left 3 per cent are destroyed by *Pteromalus* parasite as pupæ; of the 457 left 93 per cent are destroyed by birds—leaving 32 to complete the life cycle.[1]

A few cases are known in which a plant has a poisonous effect on soil pests in its vicinity. Thus, wireworms will not affect some varieties of flax, and it has been found that this is due to the excretion by

[1] A. D. Imms, *Insect Natural History*. Estimate by J. E. Moss. 1933.

the roots of prussic acid. Beans actually appear to kill wireworms, their numbers have been known to drop by three-quarters after this crop.[1] While it is not a practical proposition to breed poisonous food plants, it may be possible to breed plants with those parts poisonous which are not used for food. This already occurs, for example, in the potato, the tops containing a poisonous alkaloid.

Probably the most devastating of diseases are the moulds, mildews and rusts—mainly microscopic fungi with remarkable powers of reproduction. A grain of wheat infected with bunt is estimated to produce 12 million spores. It seems that these things are particularly liable to attack a plant when a spell of damp cool weather follows a period of drought. This observation provides us with a possible clue. Most of these fungi spread by means of spores which germinate after a period of twelve hours or less of damp atmosphere and gain entrance by the stomata or breathing pores. These pores close at night, and also by day if the plant is losing moisture or the day is dull. These stomata consist of two semicircular cells, end to end, filled with sap and working as one unit, looking somewhat like the inner tube of a motor tyre. If this tyre tube is stood upright and gradually deflated, there will come a point when the top half collapses on to the bottom half, closing the circle. This, roughly, is how the stomata work: when deflated by lack of water they close. Or, when the sugar formed during the day by photosynthesis is all converted to starch by evening, the cells of the tube can no longer maintain their 'turgescence' or rigidity (by 'osmosis'). During dry weather, however, the plant has to maintain a comparatively strong solution of sugar by night and day in order to extract moisture from the soil (again by osmosis). Along comes a drizzly day, giving a saturated atmosphere but leaving the plant roots dry. (It takes an inch or two of rain to wet a foot of bone-dry soil.) The stomata cannot close easily, on account of the increased turgescence of the plant owing to reduced evaporation and the high sugar concentration. Such weather, however, is ideal for the production and germination of fungus spores. The fungal shoots creep through the stomata and once inside they have a permanently damp atmosphere in which to thrive. Support for this view is contained in the recent finding that lettuce with a high sugar content are susceptible to botrytis.[2] In a comparison between rows of cloched lettuce I found the incidence of mildew proportionate to the length of time the soil had been under organic treatment. With one

[1] A. C. Evans, *Ann. Appl. Biol.*, 1944, 31, 235.
[2] F. W. Went, Internat. Hort. Congress, 1952.

year of compost applications there were about 85 per cent affected. With two years about 35 per cent, and three years only 5 per cent.

A shortage of water can also be caused by a plant being grown under bad soil conditions. If, for some reason, the roots have been killed, or unable to develop—as, for instance, through waterlogging or panned ground, or an excess of soluble salts, or even pests eating the roots—this will occur.

Mildews are much more serious on plants that are grown under conditions short of silica.[1] This may have some bearing on benefits ascribed to silica by the bio-dynamic school.

Mildew in cereals is very much worse in spring-sown crops than in the autumn-sown under present-day methods.[2] This, no doubt, is due to the rapid drying out of the soil after spring cultivations—or perhaps the plant develops too late to take any benefit from the organic matter left by the residues of the previous crop.

There is a very definite connection between nitrogenous manure and incidence of mildews on cereals as shown by Rothamsted.[3] The least affected plots were those receiving dung or phosphate or potash only, or no manure at all. The worst were those receiving 4 or 6 cwt. of sulphate of ammonia or 5 cwt. of sodium nitrate as a spring application. A summer fallow, which would increase the nitrogen content of the soil, also greatly increased susceptibility. While in pot experiments plants became susceptible ten days after the addition of nitrogen and the susceptibility increased with the dose. It seems as though an excess of nitrogen is aided and abetted by a shortage of decomposable organic matter—there would be little in the fallow or in the nitrogenous manure plots. As shown before, a good deal of the easily available nitrogen in the dung will have been lost during the winter, while the remainder will be held in a gradually available form. A plant overfed with easily available nitrogen has softer tissues, on account of the lower proportion of carbohydrates in the plant, and its cell walls will be more quickly lysed (dissolved) or ruptured by a fungus during the comparatively short periods when atmospheric conditions allow it to multiply. This also may be the reason for the prevalence of blackleg in sugar beet (*Phoma Betae*) where sulphate of ammonia is used.[4]

Chocolate spot in beans has caused this crop to be given up in

---

[1] F. Wagner, Phytopath Ztschr., 1940, 12, 427.
[2] Rothamsted Expt. Sta. Rep., 1952, 91.
[3] Ibid., 1951, 85.
[4] Rothamsted Expt. Sta. Rep., 1947, 75.

most parts of the country, especially the wetter districts. This is a crop, however, with which organic farmers do particularly well: there seems to be a very strong connection with the manuring and the incidence of the disease.

It is claimed by the advocates of artificial manuring that one must use adequate amounts of potash to balance these ill effects of nitrogen. So far as my own experience goes, it has not given any noticeable benefit.

There is at least one disease, 'take all' of wheat,[1] which is apparently improved by sulphate of ammonia and other nitrogenous manures in moderate amounts. This treatment does not cure the disease in any way, but enables the plant to grow fresh tillers and roots faster than the fungus can invade them. It seems rather to be a reflection of the poor nutrient status of the soil under these conditions (see also page 200). Another disease, 'eye spot', was increased by harvest time by the use of sulphate of ammonia, though the yield of grain had increased above the control owing to the much larger plant. This, too, appears to be largely a case of shortage of nutrients in the soil.

Another possible means by which plants resist disease is that they may absorb antibiotic substances from their associated mycorrhiza, if any, or from the fungal and bacterial population surrounding the roots of the plant.[2]

It is estimated that about half the known fungi and actinomycetes produce these antibiotics, which raises an interesting evolutionary point in connection with mycorrhiza. Those plants infected by a type of fungus producing an antibiotic effective against one of the plant's diseases, may, while apparently not benefiting the plant in any way nutritionally, still ensure its survival over its neighbours. Thus a strain of the species bearing mycorrhiza will develop and eventually occlude the other types. This may, in some instances at least, be the origin of the habit.

An interesting article in this aspect of disease resistance occurred in *Nature* of 17th April 1948, entitled 'Production of an Antibiotic Substance on Wheat Straw and other Organic Materials and in the Soil', by E. Grossbard of Cheshunt Research Station. A number of workers quoted had reported that organic manuring reduced the incidence of certain soil-borne diseases, and it was suggested that

---

[1] M. D. Glynne, *Ann. Appl. Biol.*, 1951, 38, 665–88
[2] For a comprehensive list of fungal antagonisms, see S. A. Waksman, *Soil Microbiology*, 1952, 278.

antagonistic microbes were the cause. It was found at Cheshunt that a number of organisms, including *Aspergillus clavatus, Aspergillus terreus, Penicillium patulum* and *Streptomyces antibioticus* formed antibiotic substances on moistened sterilized wheat straw in the laboratory.

Penicillium patulum would produce an antibiotic substance on composted wheat straw only in the presence of glucose. This limitation also occurred to some extent with fescue, lucerne, mustard and sainfoin straws. It seems likely then that fresh material, which would contain a proportion of sugars, would be more suitable a medium than the composted, as composted material has nearly all of its easily available, energy-rich material destroyed.

It was found that antibiotic substances in soil, too, were only present in appreciable quantities when it contained a proportion of carbohydrate, such as fresh straw or glucose.

Some work is being done on the subject by Rothamsted[1] also, where it was found that an actinomycete has an inhibitory effect on Brown Foot Rot of wheat (*Fusarium Culmorum*). The inhibitory substance secreted by the actinomycete is said to be absorbed by the soil minerals and organic matter under laboratory conditions and thereby rendered ineffective. These substances would only act in the immediate vicinity of the microbe that produced them: thus the obvious path by which such substances would be absorbed by the plant is from the enveloping sheath of micro-organisms around the roots. These organisms appear to feed on the dead surface cells and root hairs and probably on the proportion of cell sap which diffuses outwards. Thus the absorbing power of various soil components would not play a great part in neutralizing the beneficial effects of antibiotics under natural conditions. Neither is it likely that antibiotic excreting organisms would ever gain the upper hand under natural conditions—they are more likely to keep one another balanced. Neither could disease organisms develop to excess.

Cases have been known under controlled conditions where the introduction of a fungus to prey upon a pathogenic type has been successful. Take R. Weindling's[2] experiment, in which a soil fungus, *Trichoderma Lignorum*, controlled the damping off of citrus seedlings, and another by A. Lal,[3] showing that it controlled 'take all' in wheat and barley. A similar control has been recorded of the Panama dis-

[1] Rothamsted Expt. Sta. Rep., 1951, 56.
[2] Phytopath., 1932, 22, 837; 1934, 24, 1153; 1936, 26, 1068.
[3] *Ann. Appl. Biol.*, 1939, 26, 247.

ease of bananas by a soil actinomycete. It seems probable, then, that a high soil population will contain some types that can control certain pathogens, and a well-fed soil with its organic matter at the surface provides the best opportunity for their increase.

Another interesting example[1] has been noted in which green algae aerated the solution in which tobacco plants were growing and increased the immunity of the roots to fungal decay.

One of these antagonisms has been fairly well investigated by F. A. Skinner at Rothamsted.[2] An antibiotic producing actinomycete, *Streptomyces albidoflavus*, has been shown to compete with the previously mentioned fungus, *Fusarium culmorum*, in three ways:

(a) By direct attack on the fungal hyphae,

(b) By antibiotic action,

(c) By competition for nutrients.

These results were achieved under laboratory conditions, so that it is hard to say which effect is the most important in the soil. It is sufficient, however, that they do happen.

An example of the effectiveness of these antibiotics is shown by the fact that milk from a cow treated with penicillin was mixed with that from 200 other cows and the whole batch was spoiled for cheese-making. The cheese-forming bacteria were prevented from multiplying.

It must be borne in mind that though the introduction of one organism to prey upon or antagonize a pathogenic species may be successful under laboratory conditions, where a sterilized medium is generally used, to add such an organism to the soil is generally futile. Before the added organism can compete with the organisms already present, conditions must be arranged to suit it, particularly as regards its food supply. Parasitic or pathogenic organisms have a source of food already in their host; the saprophytic types must have their food provided, too, in the form of organic matter, generally as fresh plant or animal wastes, which are high in energy providers. Pathogenic organisms, then, will tend to thrive where organic matter is low. If only for this reason—their competitors cannot exist.

This is not to say, of course, that organic methods will render crops immune to all pests and diseases, particularly under the highly artificial conditions (by comparison with their natural habitat) that they are grown. It is often said that organic methods render a potato plant immune to blight. This is probably true under the drier conditions

---

[1] H. D. Engle and J. E. McMuntrey, *Journal Agric. Res.*, 1940, 60, 487.
[2] Rothamsted Expt. Sta. Rep., 1952, 59.

usually found in the east of England and other dry climates. Indeed, it has been found that spraying is unprofitable in such areas.[1] Given the high rainfall of the west of Britain, coupled with several days of sea-fog and cool conditions, it is difficult to see how a sub-tropical plant, totally unfitted to such conditions, can avoid falling victim to disease. If its original environment precluded the conditions necessary for the development of the disease, at least during its growing season, the plant obviously will not have evolved a resistance to it.

It has been shown that the original prototypes of our potato, *Solanum andigeum* and *tuberosum*, coming from the hot climate of Chile, have no innate resistance whatever to blight. (Blight spores do not germinate if the temperature is over 72° F.) Neither have many other wild varieties—only two, so far, showing any resistance. One of these, *Solanum demissum*, is being used to develop resistant varieties by hybridization with our present varieties—with some success, as in the new variety Orion. Two more varieties—Canso and Keswick— have been developed in Canada, but on trial in this country Canso developed the disease through a particular strain of blight.

One might give a similar instance of this with our apples, which suffer from scab disease. The parents of most of our best apples are derived ultimately from eastern Europe, where the climate is dry. It is noteworthy that our own wild crab apple and those allied varieties, the codlins, usually have a high degree of resistance. The same occurs in the case of the peach and peach leaf curl, and doubtless many others. Sir Albert Howard quotes the incidence of rust in wheat, which is worse in hot climates than in its native temperate climate.

However, even in these cases, there is little doubt, from experience, that organic methods do lend a considerable degree of resistance over the orthodox, probably due to stronger and thicker cell walls.

Eelworms are an increasing menace these days, particularly on the potato (costing us 250,000 tons annually), tomato, and sugar beet, amongst food crops, and to a minor extent in practically all cultivated plants. These animals, which are commonly found attacking wild plants in nature, are, of course, controlled by their predators and scarcity of food under such conditions. Under cultivated conditions, however, they are given ample opportunities for increase and dissemination. Their predators, on the other hand, are largely suppressed. No doubt, bacteria and protoza play a part, but it is known that some twenty-five species of fungi enmesh them in their mycelium

[1] A. Beaumont, J. H. Bant and I. F. Storey, 'Potato Spraying Trials in Yorkshire', *Plant Path.*, 1953, 2, 2, 56.

('fungal threads') and digest them. A species of Arthrobotrys has been found attacking nematodes on the roots of Sitka spruce, with two more species which could not be grown on an artificial medium.[1] It seems one even has a resting stage inside the potato eelworm cyst,[2] while another species, attacking the stem eelworm (*Ditylenchus dipsaci*), produces sticky spores which bore into the eelworm on germination.[3] These fungi require supplies of easily decomposed organic matter. That this provides an effective control of eelworm, under some conditions, has been shown by Mr. Timpson, Government Agriculturist in S. Rhodesia, particularly in the case of tobacco, but also for garden vegetables. An account is published in the *Rhodesian Herald* for 7th July 1944, and repeated in Sir Albert Howard's *Farming and Gardening for Health and Disease*. The possibility of such control in this country is being investigated by the Agricultural Research Council.

Some work is being done in Rothamsted[4] on the potato eelworm, using a heavy clay soil, modified in different instances by the use of sand, peat, artificials and compost, in large pots. The last is presumably made with synthetic nitrogen and is perhaps the reason why it did not give any control. Or, more likely, it was too well rotted to provide the fungi with food. The eelworms in the first year greatly increased, showing a 35-fold increase in eggs. Sand gave a slight reduction and peat a slight increase in numbers. The potatoes grew best where artificials or compost were added.

It was found also that eelworms did not move far from their original point of infection, 80 per cent remaining within 2 in. of it, though 1 per cent moved as far as 6 in. It is not unlikely that shallow cultivation would do much to prevent its movement to the lower and more compact layers of soil. This would have the double effect of confining them to the top layer, leaving at least part of the soil clean for roots, and keeping the eelworms where there is the greatest fungal activity. It was shown, too, that increase in the cooler, deeper layers of soil was slight, only half that at the surface (using well-mixed soil in boxes).

Previous work showed that certain grasses stimulate the eelworm cysts to hatch out,[5] when, of course, in the absence of the potato, they

[1] Rothamsted Expt. Sta. Rep., 1950, 84.

[2] Ibid., 1949, 75.

[3] J. B. Goodey, Trans. Brit. Mycol. Soc., 1951, 34, 270–72.

[4] Rothamsted Expt. Sta. Rep., 1950, 85; 1951, 96.

[5] B. G. Peters, 'Review of Work on Potato Root Eeelworm', Rothamsted Expt. Sta. Rep., 1950, 147.

die. The Meadow Grasses (*Poa trivialis* and *pratensis*) were most effective. Ryegrass and maize were effective to a moderate extent, and slightly in the case of Cocksfoot and Foxtail (*Alopecurus pratensis*). It seems that the use of leys containing these grasses would go some way to controlling these pests. Work on fourteen different grasses is now in progress. The presence of mustard plants in the crop sometimes inhibits the hatching of eggs.

They have also started work on a soil amoebae which consumes this eelworm. This organism was discovered in Holland.

I have done an experiment myself, growing potatoes on adjacent plots with compost or artificials over two successive seasons, one plot shallowly cultivated and the other dug. The standard market-garden quantities of compost and artificials were used, the potatoes being laid on the compost and earthed over on one plot, the artificials spread over the ridges before earthing on the other. A handful of eelworm-infected soil was piled over each potato to ensure infection. All the potatoes appeared to be slightly affected the first season, but it was interesting to note that the eelworms did not travel through the compost but remained immediately around the potato.

The second season showed a marked difference. The composted potatoes were about a fortnight behind the artificials in development, as is commonly noted, but eventually made larger growth and remained green for about three weeks longer than the artificials' plot. The artificials' potatoes suddenly turned yellow when they reached full size and quickly died off. Eelworm counts later in the season showed six times as many eelworms in the artificials plot, (134 against 22 per 100 gm.), while the compost plot gave 50 per cent greater yield. Oddly enough, a further count in November showed a reduction in numbers to 26 and 19 respectively. It seems as though it was not only the potatoes that were more robust on the compost plot.

There seems to be a considerable difference between the organically activated compost and that used by Rothamsted.

Attempts to destroy the eelworms by soil disinfectants have usually shown a final increase in eelworm numbers, possibly by destroying the fungi and other enemies, but more often by sterilizing the soil, releasing nitrogen, giving a larger plant for the eelworms to breed on.

Plant viruses are one of the most difficult and elusive of the maladies with which we have to deal, mainly on account of their small size and because it has been difficult to find materials which will check them without harming the plant. Their study, too, has been complicated by one virus giving different symptoms in different plants, while similar symptoms have been due to different viruses. Furthermore, one kind of virus may have different strains which may affect one plant yet not another, while both affect a third. It is only recently that some sort of intelligible background is being formed from a very involved subject.[1]

Here again we have to thank Rothamsted for some interesting observations, especially Messrs. Bawden and Pirie, whose painstaking work on a complex subject is worthy of admiration.

Two substances have been extracted from a fungus, *Trichothecium roseum*, with very different chemical properties, but with the common property of inhibiting the spread of virus disease in beans and tobacco plants, when sprayed on or injected into the leaves.[2] One of these substances is also antifungal but, unfortunately, damages the plant as well. The other substance has no harmful effects but only lasts for a day or two. There is, here, another way by which a mycorrhizal fungus may provide the plant with preventative substances. Obviously, a mycorrhizal fungus does not give the plant any substances toxic to it but it would quite easily give materials toxic to some invading organisms, if a virus can be so called.

Another interesting observation from the same source is that the sap of one plant can sometimes inhibit virus when applied to another species. Sugar beet, cucumber and thorn-apple juice was used on tobacco plants. This may account, in part, for the observed freedom from virus in the case of strawberries kept under rather weedy conditions. It is usually found that stragglers in the hedge, though obviously suffering from competition, nevertheless appear to be healthy. It may also, in part, account for the observed benefits, in some cases, of mixed cropping. The closely intermingled roots, and possibly leaves, may be subjected to some interchange of sap. It has been shown that some virus of tomatoes and potatoes can be transmitted from root to root in soil and culture solutions by contact,[3] so it seems that interchange of sap also is not impossible. These possibilities are, of course, in addition to the fact that aphids, the carriers of virus, would have rather more difficulty in finding their host plants under such conditions.

However, the evidence that organic methods will prove an infallible cure for virus diseases is not convincing and some of the claims

---

[1] 'Symposium on the Nature of Virus Multiplication', *Soc. Gen. Microbiol.*, 1952.

[2] F. C. Bawden and G. G. Freeman, *J. Gen. Microbiol.*, 1952, 7, 154.

[3] M. L. Roberts, *Ann. Appl. Biol.*, 1950, 37, 385–96.

of organic enthusiasts must be taken with reserve. I have myself kept a strawberry plant (Cambridge 582) infected with crinkle virus on the site of an old compost heap for three years. It certainly looks greener, but remains stunted and unfruitful. Various other sites were tried, including hedge-banks and woodland, but none recovered and some died.

The reason why apparent recovery has been noted in some cases seems to be this. According to research by East Malling most, of not all, strawberry (and some other) plants are infected with some viruses permanently. All stocks of Royal Sovereign Strawberry, for instance, are infected with crinkle virus. This, however, does not affect this particular variety unless infected by a further kind of virus; or, and this is the important point, it is grown under bad conditions. Improvement of soil conditions or plant nutrition will, under those circumstances, restore the plant to apparent health. A tomato virus is similarly 'masked'. This, unfortunately, does not apply to all varieties of plants or viruses.

Wild strawberries are not found infected with virus—all viruses seem lethal to this species—but virus-infected wild blackberries and raspberries may be quite commonly observed. These occur far from human habitation, thus eliminating the possibility of garden escapes. Presumably, these stocks die out eventually and their place is taken by more vigorous seedlings—thus ensuring that even the vegetatively propagated species are reproduced sexually occasionally. Even so, one does sometimes come across examples of cultivated varieties in isolation remaining free from virus and disease—while in a neglected condition—for many years.[1,2]

There may be a direct connection between manurial methods and the severity of virus.[3] It has been found that additional phosphates have increased the amount of tobacco mosaic virus in the plant—in some cases up to forty times, though part of the increase was due to increased growth with the additional nutrient. The concentration in the sap doubled, though only about a third of the virus particles were found in the sap, most apparently remaining in the tissues. It was concluded that virus production occurred at the expense of normal plant proteins, and the more phosphorus added the greater the proportion of the proteins converted to virus.

However, a similar test on a potato virus produced inconclusive

[1] *Commercial Grower*, 9th July 1953.
[2] *N.A.A.S. Quart. Rev.*, 1953, 21, 46.
[3] F. L. Bawden and B. Kassanis, *Ann. Appl. Biol.*, 1950, 37, 138.

results, another example of the extraordinary complexity of the subject.

Plants growing in shaded or excessively warm conditions may also show a greater concentration of virus in some cases. Though those conditions increase the number of virus particles present, they do not increase its power to infect a fresh host—except in so far as a sucking insect may transmit a larger dose of virus to the next plant.

Some viruses have an almost benign influence by preventing infection of the plant by a more severe form.

Organic methods do undoubtedly lead to a very much lower incidence of virus. From my own experience I can say that there has been no tomato virus on my holding for years and very little lettuce or potato virus. Blackcurrant reversion has disappeared (and 'big bud' is seldom seen) and I have never seen any virus diseases of brassicas. The reason is probably the almost complete extinction of sucking insects, especially aphids. Strawberries, unfortunately, still show virus at times—but this may be due to soil conditions.

Interesting support[1] is provided by the observation on potatoes of the incidence of virus. The greatest amount of virus infection of the two viruses, leaf roll and rugose mosaic, was found on potatoes grown on dung; sulphate of ammonia increased leaf roll and sulphate of potash increased rugose mosaic. The greatest aphid populations were found on plants treated with dung, sulphate of ammonia and superphosphate—but, oddly enough, they were reduced by sulphate of potash. It seems that nitrogenous manures at any rate had detrimental effects.

A soil difficulty examined by Rothamsted for the last fifteen years is that of 'clover sickness'.[2] Clovers seem unwilling to grow on the same ground repeatedly, even when other crops are occasionally grown in between, unless fairly large amounts of farmyard manure are added. They found that by heating the soil to 180° F. for two hours, but only if moist, this effect could be eliminated, though it quickly returned; but it was only slightly affected by chemical sterilization; not at all by leaching out the soil with water, or by oxidization (aeration). They conclude that there is some toxic substance exuded by the roots. It was also found that this depressing effect extended to some other plants—notably lettuce, radishes and spinach beet, but not to grasses or cereals. The notable part about this experi-

---

[1] L. Broadbent, P. H. Gregory and T. W. Tinsley, *Ann. Appl. Biol.*, 1952, 39, 509.
[2] Rothamsted Expt. Sta. Rep., 1951, 152.

ment is that farmyard manure, and to some extent other, mineral, substances of colloidal nature, reduced this toxic condition apparently by absorption.

It seems likely that the destruction of this toxin (it might even be produced by an associated soil organism) is accomplished by certain bacteria living on the soil organic matter. These increase rapidly after heat sterilization (much more so than after chemical sterilization) owing to the large amount of dead organic material and nitrogen released and freedom from competition. This phase soon passes with the reincorporating of those materials into the soil organisms. This rapid increase also occurs when farmyard manure is added to the soil, and lasts much longer. It is notable that the length of time during which the manure is effective depends upon the quantity applied.

The lodging of cereals, while not a disease, in the sense that it does not necessarily depend on the attack of pathogenic organisms, is seemingly connected with the nourishment of the plant. Rothamsted in the wet summer of 1950 reported extensive lodging of their cereals. Friend Sykes, of Chantry, claims not to suffer any during wet seasons. This was not due to a starved condition of his wheat, as shown by both his yields and protein content, which were well above average. Most organic farmers report similar findings.

Another type of 'lodging' is connected with various diseases. The disease 'take all' of wheat is a fungus which attacks the root and crown of the plant and overwinters on the stubble. It can be well controlled by the undersowing of trefoil (*Medicago lupulina*). This hastens the decomposition of the straw in the late summer by keeping it moist, so that the fungus has nothing to live on. It has also been found that depriving the straw and soil of nitrogen, or competition for nitrogen by other organisms, or good aeration, soon kills out the fungus. The latter conditions occur under a leafy cover, but not if the stubble is ploughed in. The explanation given by Rothamsted is that the trefoil takes up the nitrogen from the soil and starves the fungus,[1] but in view of the habit of legumes to shed their nodules and release nitrogen at the end of the season this is difficult to understand. In any case, fungi generally are much better extractors of nitrogen than higher plants. Another worker, Fellows, of Kansas, found that the disease could be controlled with poultry and horse manure—both rich in nitrogen—and thus able to rot the straw quickly.

It is noteworthy, too, that when an attempt was made to germinate

---

[1] S. D. Garret and H. H. Mann, *Ann. Appl. Biol*, 1948, 35, 435.

the aerial spores of this fungus on wheat roots it was found that it would only take place in a sterile medium as the competition from the microbial population was too fierce. A fertile soil, then, would ensure freedom from this mode of attack also.

Another disease of wheat, Brown Footrot, is only serious where poor soil conditions obtain. It is a normal colonizer of cereal stubble, but if the soil becomes acid or waterlogged or the plant is starved it will invade the living roots. A root disease of cotton has been shown to make an attack only if the carbohydrate content was low, so that shaded or overcrowded plants, or plants overfed with nitrogen, are likely to be attacked.[1]

These last few diseases, it seems, are directly or indirectly a result of poor growing conditions.

Lack of aeration of the soil, by waterlogging or compaction, of course, can kill roots directly or by so weakening their metabolic processes that they are unable to resist invasion by weakly pathogenic organisms. This is often the case with strawberries.[2]

Laboratory work on the fungus disease Violet Root Rot, that attacks grasses, clovers and vegetables, has shown that in a soil rich in carbohydrate but poor in nitrogen this disease forms resting stages, while if plenty of nitrogen is provided the disease continues to vegetate until the ready available carbohydrate is used up—when the fungus dies. Perhaps in this instance, as with 'take all', the use of dung, a leguminous plant, sewage or urine, as a direct application to the soil, when the crop is cleared, would go some way towards controlling it. But, as always, one must be cautious about taking laboratory evidence too literally.

A fungus disease of brassicas called Ringspot (distinct from the virus of the same name) is very prevalent in western coastal regions of the British Isles, largely, no doubt, on account of the mild humid conditions in autumn. Sprouts and broccoli grown on a demonstration plot by the local N.A.A.S. were affected progressively, year by year, until, after five years, the plots had to be abandoned. I, too, had the disease severely, and exhortations to apply potash had no effect. Since the use of artificial fertilizers and other readily soluble forms of nitrogen was abandoned these crops have steadily improved and the disease no longer causes trouble.

Soil acidity has some influence on the incidence or degree of some diseases. Club root is associated mainly with acid soils, while Potato

---

[1] F. M. Eaton and N. E. Rigler, *J. Agric. Res.*, 1946, 72, 147.
[2] W. P. Cheal, Botley Fruit Sta. Rep., 1952.

Scab and 'take all' of wheat are more often found in chalky soils. A high level of organic matter helps to ameliorate such conditions.

The most effective way of controlling a pest or disease is by introducing a parasite to feed on it. There are numerous examples of this. The white fly of greenhouses is controlled in many nurseries by the introduction of a minute, parasitic fly. Similarly, one can use a parasitic insect to control woolly aphides on apple trees. An eelworm is known to parasitize a species of thrips.[1] The introduction of the Giant Toad to eat the white grub of sugar cane is a classic. Dr. Hugh Nicol has written a very interesting 'Pelican' book, *The Biological Control of Insects* (and a few other things). He mentions a parasite introduced to Canada from France to clear up the Codlin Moth of apples. I have never heard of it being introduced to this country, but it seems possible, from the description, that our Codlin Moth is a different creature.

One of the most remarkable histories of insect interrelationships is related in the Unilever quarterly, *Progress*, by Mr. R. Leach. It appears that a sucking bug damaged embryo coco-nuts and caused them to drop off. It had also been noted that a large yellow ant, found nesting in the tree-tops, could control these bugs. Another (small black) ant lived amongst the aerial roots at the foot of the trees, while yet another (small brown) ant made covered runways all over the trunks.

These two small species, however, did nothing to control the sucking bug—in fact, it was found their presence encouraged it. They would attack the large yellow ants in hordes and drag them into cracks in the bark. It was found, further, that if a 'bridge' of foliage or creeper enabled the large yellow ant to bypass the nest of black ants at the foot of the tree, they would continue to control the sucking bugs. But this did not occur where the small brown ant colonized the tree bark. But, it seems, these two small species, brown and black, were antagonistic to one another. So, in order to keep the bugs under control, large yellow ants were needed, which in turn needed a colony of black ants to eat the brown ants and a bridge to dodge the black ants.

Diseases, too, may well have their predators. There is recorded a case of two gardeners whose garden was afflicted with club root.[2] Gathering all the club-rooted material they could find from their neighbours, they composted it and used it in the garden. Since when

[1] A. M. Lysaght, *J. Anim. Ecol.*, 6, 169–92.
[2] *Soil Association Journal*, Vol. 2, No. 2.

they have had no more trouble. It is not unlikely that such a collection of diseased material would prove particularly attractive to the development of organisms which could feed upon the pathogenic fungus concerned, and when spread upon the ground continued the good work on the diseased roots and fungus in the soil. In pot experiments at Rothamsted it was found that the number of diseased plants in club root infected soil dropped to 19 per cent after growing ryegrass, 28 per cent after cabbage—which, curiously, is a susceptible plant—and 57 per cent after leaving it fallow. So it seems better to grow a susceptible crop than leave the soil idle. They found, however, that results were not consistent, so there are probably other factors involved—including the fact that there are different strains of the fungus. A further observation was that hatching of resting spores (oospores) was delayed in a dry or alkaline soil, but was hastened in acid or wet soil. So it seems that the practice of liming does not cure the disease but merely delays its onset.

It is of interest that a number of non-cruciferous plants can also be affected by this disease, including nasturtium,[1] poppy, mignonette, ryegrass and cocksfoot grass, and the disease can spread amongst them by means of swimming zoospores. These, however, are short-lived in the absence of a host. The longer-lived oospores do not seem to be formed on these plants. This may be the reason why growing them in rotation helps to decrease the disease—the oospores hatch but cannot form any further resting bodies.

Several have testified to the qualities of composted diseased tomato haulms.

A few words on insecticides and fungicides are appropriate here. I do not think that under the conditions of specialization made necessary by economic conditions we can avoid the use of these all the time.

There are preventative measures that could be taken in particular instances but it is not economic to practise them—to take a rather extreme example, the hand-picking of caterpillars. The occasions when preventative and curative measures are necessary under organic methods are rare. Their use is more commonly a prop to bad husbandry.

An insecticide or fungicide in its ideal form should be effective on the particular complaint in hand and nothing else. Few, if any, come up to this standard, for obvious reasons.

Secondly, its toxicity should disappear as soon as it has completed

---

[1] Rothamsted Expt. Sta. Rep., 1948, 55.

its task. In effect this means that it should break down into harmless compounds or be broken down by soil organisms.

Taking this view, poisons of vegetable origin, nicotine, derris, quassia, pyrethrum and the like, will be safe to use, as by their nature they will be broken down by soil organisms. If this were not so, large areas of the world would now be sterile in those parts where the plants producing these poisons grow.

It is true that some of these substances are now made synthetically, but this raises a debatable point, in which there seems to be no evidence to support any contentions on either hand.

Thirdly, if there is any danger of the material harming soil organisms, the minimum quantity of wash compatible with the job in hand should be used. In this respect the use of the low-volume, atomizing, spraying machines seems preferable.

A case can be made for the use of some of the synthetic organic phosphorus compounds. (Organic in the chemical sense.) Those known briefly as HETP and TEPP break down in twenty-four hours or so to alcohol and phosphoric acid, which are substances common in Nature. Some of these phosphorous insecticides, however, are considerably slower in their rate of decomposition and are absorbed by the sap and if there is a risk of much spray falling on the soil it seems better to avoid them. Their use on food crops within six weeks or so of harvesting is inexcusable, as the manufacturers do, in fact, point out. Rothamsted experiments have shown that one of these insecticides, Parathion, Schradan, or E605, may be found in honey, having been gathered by bees in contaminated nectar.[1] Such insecticides should be reserved for special cases, as, for instance, the preservation of strawberry and other stock plants (i.e. non-fruiting) against the spread of virus diseases by aphides.

There is another class of synthetic insecticides, the chlorinated hydrocarbons, including DDT and BHC, which many regard as one of the most deplorable inventions ever released on the earth by mankind. Much disturbing evidence has accumulated in the United States, and the position is well summed up by Martin S. Biskind, M.D., in his statement presented before the Select Committee to Investigate the Use of Chemicals in Food Products, House of Representatives, U.S., 12th December 1950.[2]

'Somehow a fantastic myth of human invulnerability has grown up with reference to the use of these substances. Because their effects are

---

[1] Rothamsted Expt. Sta. Rep., 1952, 111.
[2] Reprint issued by the Soil Association.

cumulative and may be insidious and because they resemble those of so many other conditions, physicians for the most part have been unaware of the danger. Elsewhere, the evidence has been treated with disbelief, ignored, misinterpreted, distorted, suppressed or subjected to some of the fanciest double-talk ever perpetrated.'

The dangers of these substances are threefold—they are persistent in the soil, they are absorbed by the plant and they accumulate in the fat of animals and human beings.

Evidence supporting the accumulation of DDT in the fat of animals is abundant, but the extent to which human beings are accumulating it does not appear to have received any attention in this country, so far as my inquiries show. America, however, is more alive to the danger. Texas Research Foundation[1] acquired samples of meat and milk foodstuffs from various retail sources. They found that all the samples contained DDT, varying in amounts from 3·10 parts per million in lean meat to 68·55 p.p.m. in fat meat, and from 0·5 p.p.m. to 13·83 p.p.m. in milk. When one considers that 5 parts per million are considered to be toxic, it seems, as they rightly point out, that there is a very real hazard in its use.

According to Lord Douglas of Barloch, quoting an experiment in a debate on Chemicals and Food Supplies, 4th July 1951, rats fed with 1 per cent carbolic acid in their food did fairly well, but when fed DDT, as 1 part per million, they died—through its accumulation. One part per million is about 1 teaspoonful in 10 tons of food.

It is, true, however, that there is some contrary evidence on this point. In an experiment by other workers, rats were fed 350 p.p.m. of DDT in their food for sixty weeks without any toxic changes or accumulation in the body. It seems that there is a defect in one or other of these experiments. Perhaps certain foodstuffs have a protective influence, or there were no fats in their food, or the strain of rats was resistant to DDT. Or even perhaps because DDT sometimes contains impurities which are toxic or protective to rats.

The process by which human beings absorb these substances is somewhat as follows. The plant absorbs spray residues either directly through its surface or from the soil. This in turn is eaten by the human being or a farm animal. The DDT ingested passes into the blood-stream and is deposited in the fat. In the case of the animal fed on contaminated food, it is passed to the human being through its meat or milk.

Dr. Biskind records a case of a patient nursing a baby, whose milk

[1] *Soil Ass. Journal*, Vol. 6, No. 1.

at first contained 118 parts per million of DDT and gradually fell to 5 p.p.m. After seven weeks there was still 8 p.p.m., when feeding of the baby was discontinued for obvious reasons. It is significant that the mother's health improved considerably by being relieved of the load of DDT. A number of other such cases are recorded.

Cows' milk, according to reports from the U.S. Department of Agriculture. contained 0·5–25·0 p.p.m., and in one case the butter-fat contained 259 p.p.m. Even where DDT is not used on the farm, bought feeding stuffs contain it, as shown by the cream of one dairy giving 4 p.p.m., though none was ever used on the farm supplying it.

My own experience, while not conclusive, is highly suggestive. I know from experience of using DDT some years ago that I have become very sensitive to it, and this, no doubt, accounts in part for my extreme distrust of it. It is significant that over the past three or four years, when the use of this material has become increasingly common for sheep dips, wireworm and caterpillar controls, and use about the farmyard for fly control, both my wife and I have become increasingly liable to digestive upsets due chiefly to fat. Not, curiously enough, from fat in general or from particular kinds of fat, but from particular samples of fats. For instance, butter, bacon or meat may be quite innocuous on one occasion but produce a severe reaction the next, invariably repeated with the same sample. Further, woollen socks—of recent manufacture—often cause skin irritation. This effect can, however, be removed by soaking in a tank of paraffin. (DDT is soluble in fats and oils but not in water.)

It is interesting to note in this connection that the recently published (Oct. 1953) Report of the Working Party on Toxic Chemicals in Agriculture (Residues in Food), while stating that they were unable to find any evidence of poisoning by spray residues in Britain, made it clear that there was no official body charged with the search for such evidence. It is, in fact, nobody's job to find out whether we are being poisoned or not, so we just don't know.

Not being in a position to analyse food and other materials myself, I have sought for any evidence that food in this country has been subjected to analyses for DDT, but have found none, or, indeed, any evidence that the position is even appreciated.

Fortunately, in America some health authorities are alive to the danger—that of Chicago, for instance, has banned its use by suppliers of milk to the city. 'Effective immediately, the use of DDT in any form and for any purpose whatever, in connection with milk production, transportation, processing, and distributing by farms or plants holding Chicago Board of Health permits is herewith prohibited. Violation of this regulation will result in revocation of permit.' Issued by Dr. Herman N. Bundsen, president of the Chicago Board of Health.

In regard to its persistence in the soil, it has been shown that applications of 25–1,000 lb. per acre showed no reduction after four years,[1] and yields of beans and the marrow family were reduced considerably by its use. A general depression of plant growth was built up in five years, in some cases, by normal applications. BHC disappeared at the rate of 15 per cent per annum and had a drastic effect on soil flora. At 10 lb. per acre (a recommended wireworm dressing is about 7 lb. per acre), fungi and nitrifying bacteria were reduced to a very low level. Three to five years' application may produce a general depression of growth.

It seems, therefore, that we are building up a menace for perpetuity—even our composted sewage will not be safe.

BHC has a further effect on plants which, so far, has received little attention, though the scientific work on the subject is conclusive enough.[2] In experiments it caused genetic changes in onions and strawberries—the two test plants—even at the low rate of 24 oz. per acre. This resulted in stunting of the plants, and even death of the roots, where somewhat higher local concentrations occurred. In strawberries 75 per cent of the roots were found to be abnormal, in some degree, with a proprietary wireworm dust, used at the recommended rates. It was recommended that BHC should not be used on propagating stock, but the advice does not seem to have been widely heeded. This insecticide, incidentally, can actually be tasted in many crops, developing an earthy taste in potatoes, for instance. As a consequence, manufacturers have now developed a tasteless form, so that there is no longer that safeguard against eating it.

Another particularly unfortunate characteristic of these insecticides is that they destroy the pest's natural enemies, while, at the same time, the pest evolves a strain more resistant to the insecticide. The latter is probably true also of other insecticides, but does not seem to have received such attention. It is noteworthy that the use of DDT and BHC in fruit orchards was followed by a rapid build up of red spider and greenfly, in this country, and mealie bugs in citrus orchards.

Fungicides hitherto have been mainly based on sulphur and cop-

[1] C. J. Shepherd, *Rhodesian Farmer*, 1st February 1952 and 5th March 1952.
[2] M. E. Scholes, *Jour. Sci. Hort.*, 1953, 1, 49.

per, which, generally speaking, are immobilized in the soil, though in their free condition are naturally very toxic to soil organisms, and there is a risk of a build up of depressing conditions after some years. In some glasshouse soils this condition seems to have arisen already. A case has been recorded[1] where the introduction of compost to a glasshouse soil used for growing tomatoes for many years under orthodox methods failed to show any improvement—the compost remained undecayed on the surface of the soil. A possible explanation is that the soil is so full of inimical substances, from the fertilizers perhaps, but quite likely from copper or sulphur sprays, that any appreciable fungal or worm activity is impossible. The fact that the haulms from this nursery failed to rot down quickly seems to indicate this. Certainly few glasshouses can boast a worm these days—though, of course, there are other contributory factors.

Experiments are going on with fungicides of organic chemical composition, but so far they have proved expensive. Even the possibility of antibiotics is being exploited which, being substances continually produced and destroyed by soil organisms, should not cause any serious damage to them.

I have found myself that a certain amount of protection against potato blight in both tomatoes and potatoes is necessary in a cool damp climate, but, as already explained, there seems to be a good reason for this. I use a colloidal copper oil emulsion which is copper in its most finely divided and, therefore, most effective form, and the oil ensures that it sticks to the foliage better than most. The smallest quantity is used to the greatest effect. Slightly more expensive perhaps, but, taking the long view, well worth while. One or two light sprayings are sufficient—not the fortnightly drench commonly thought necessary. (Incidentally, Rothamsted in 1952 found 10 gallons of spray as effective as 40 gallons—but this may have been due to a comparatively dry season.) Considering the colossal aerial population of blight spores in this early potato district at the end of June, I think it is a remarkable tribute to organic methods that my potatoes survive at all. Potato blight can actually be smelt over the countryside on warm muggy days. Even so, a very wet summer encourages such rank growth that spraying becomes impossible and blight gets a hold.

There is a danger, too, in building a protective shield around our plants and animals. We are encouraging the less-resistant types unwittingly, with the resistant. Not only, therefore, will we not breed

[1] E. M. Walrond, *Soil Ass. Journal*, Vol. 4, No. 2.

more disease-resistant races, but we will have to continue to increase the size and thickness of our protective shield—already nearing its weighty limit in fruit growing—until, finally, in desperation, the farmer chucks the lot and goes 'organic'.

A difficulty associated with the use of pesticides and fungicides is that the pest gradually evolves a strain more resistant to the material, so that the strength has to be increased from time to time or new material found. The more frequently the material is used the more rapidly will the new strain evolve. DDT was only introduced to the public in 1945, since when pests in many places have become resistant, especially houseflies. Mr. T. Neame, a fruit farmer, told an audience a little while ago[1] that he remembered the day when $1\frac{1}{2}$ oz. of nicotine to a 100 gallons of water would control the hop aphis. Now it requires 6 oz.

It is obvious that by assiduously removing all possible traces of pests and diseases we are depriving such natural enemies as they have of means of multiplication. Once we take up the shield on behalf of the plant we must carry the whole burden. In this view a mild infestation may be more economic than the expense of 'plant protecting'.

It seems, then, that for the good of mankind generally it would be fundamentally a more economic proposition to develop varieties with a degree of resistance rather than spend much time and material on protective measures. There is, however, a snag—as long as a disease or pest exists, by natural evolutionary processes it will develop a strain which can attack the new variety. This has, in fact, happened repeatedly in the case of the greenhouse tomato disease Cladosporium, and seems to be occurring with potato blight. The different strains of apple and pear scab run into several dozens. So it seems probable that we will have a race for all time, but at least the increased resistance afforded by organic methods will give us a longer interval between the breakdown of fresh varieties.

[1] Amos Memorial Lecture 1951, E. Malling Res. Sta. Rep.

X

THE FOOD OF THE ANIMAL

As the ultimate purpose of growing most crops is to feed ourselves or our livestock, it is appropriate to quote a few words from Sir Robert McCarrison's *Nutrition and National Health*. His experiments, on a grand scale, using rats, will convince the most sceptical that health is a direct consequence of the type of food eaten. 'Certain it is that no synthetic diet that I have been able to devise has equalled in health-sustaining qualities one composed of the fresh foodstuffs as Nature provides them' (page 11).

Health, unfortunately, is a state that not many of us know, for, according to the findings of the Peckham Health Centre before the war, nine persons out of ten over 25 years of age had some physical defect. Nowadays the Ministry of Health tell us that over the past three years two-thirds of all men between 21 and 64 visit their doctors every month. Women fare even worse. It costs on the average each member of the working population £75 per annum in loss of output and medical attention. We eat 10 million aspirins a day. Such figures do not indicate health but chronic sickness. Other highly 'civilized' countries, where advanced methods of agriculture are used, the U.S.A., and to a lesser extent Holland, and Germany, can give similar figures.

It is impossible to give a cut-and-dried 'scientific' definition of health: it involves too many factors. However, the World Health Organization of the United Nations defined it very well, as follows:

'Health is a state of complete physical, mental and social well-being, and not merely the absence of infirmity.'

If we are to be healthy, then we must eat the right kinds of food and also be sure that they are of a suitable quality. Quality is difficult to define. The food may look, taste and smell good and yet not give us good health. The plant that constitutes this food may have obtained everything from the soil that it needs and yet not have absorbed, involuntarily perhaps, something needed by the animal that feeds upon it. A series of experiments by M. J. Rowlands and B. Wilkinson[1] gives evidence of this and is admirably described by Dr. Picton in *Thoughts on Feeding*. (An excellent book on nutrition.) Here is a summary of it. Mr. Rowlands noticed that his pigs did better on his own grown wheat and barley meal than bought, and wondered if it was due to the bought meal being grown with artificial manures. He therefore decided to make a laboratory test, using rats. He took clover seeds as the foodstuff, as a matter of convenience, taken from a field, half of which was manured with strawy pig manure at twenty loads to the acre, and the other half with 20 cwt. of basic slag, 3 cwt. of kainit and 1 cwt. of sulphate of ammonia per acre.

It was suspected that the trouble was connected with a lack of vitamin B. So a diet deficient in this vitamin was fed to the rats, to find out whether the clover seed would supply it—20 per cent of the ration consisted of clover seeds. One lot of rats was fed on 'dung' seeds, and another lot on 'artificial' seeds. The 'dung' seed rats grew well at first but slowed down, so the percentage of seeds was raised for both lots to 25 per cent. They then grew normally again.

The 'artificial' seed rats also grew at first and then stopped, and even after adding 25 per cent of the seeds to the diet they continued to deteriorate.

Another two lots of rats were fed for fifteen days on the vitamin B deficient diet and were by this time actually losing weight. They were then given a ration of seeds, one from the dunged soil and the other from the artificially manured soil. The 'dung' seed lot recovered, the 'artificial' seed lot continued to decline, and after another three days were obviously dying. They also were put on to 'dung' seeds and immediately began to recover.

They went further and made an extract from the dung, which they fed to the vitamin-B-starved rats. Their health improved.

Similar observations were made when comparing the effect of manuring by Sir Robert McCarrison[2] and E. Pfeiffer.[3]

It seems, then, that not only must one study the kind of food one eats or feeds to animals but also its source. This is probably one reason why analysts searching for vitamins find them so variable in

[1] *Biochem. J.*, 1930, 24, 1.
[2] *J. Med. Res.*, 1926, 14, 351.
[3] Dr. E. Pfeiffer, *Soil Fertility, Renewal and Preservation*.

quantity in different samples of the same foodstuffs. Dr. E. Pfeiffer records a carrot without any carotene—one of its most valued constituents. The point to be borne in mind is that an increasing proportion of our food, perhaps over half by now, is grown with the aid of artificial manures under conditions of low organic matter.

In this connection Dr. Cleveringa, State Agricultural Consultant for Dutch Soil Research,[1] finds that there is a relationship between poor turf and foot-and-mouth disease. He considers that a nutritional deficiency reaches a maximum in early August, when old-fashioned farmers dung their fields. These farmers are seldom afflicted with foot-and-mouth in their stock, whereas, taking the country as a whole, foot-and-mouth starts in August and gets gradually worse until the animals are stabled for the winter. He also thinks that poliomyelitis (infantile paralysis) is rooted in the same cause, the milk from deficient pastures being unable to protect children against the disease.

Organic matter is at its lowest about August and it is not inconceivable that vitamin B or similar substances manufactured by soil microbes is also at a low ebb. Even the antibiotic substances they produce may be transferred by the plant—they are certainly very effective against animal diseases.

The vitamins of the B group are manufactured by soil organisms, especially bacteria and fungi—which live on organic matter. In fact, though most of them manufacture these vitamins and other food accessories for their own use, if they do not they have to obtain them from those who do. Many of the bacteria found in soil treated with farmyard manure have to be given a yeast supplement in artificial culture to provide them with this vitamin. It is believed that most higher plants can manufacture their own, though a number of experiments quoted by Dr. E. W. Russell[2] suggest that these substances may sometimes effect yields when in the soil. However, it seems that the artificially manured clover plants of Rowlands did not contain much vitamin B. Dr. E. W. Russell, however, goes on to say: [3]"The second question—to which extravagant answers have been given by some workers—is whether the feeding value of the crop either for animals or humans is affected by the presence of farmyard manure or composts in the soil. The earlier experimental evidence on this point was conflicting, possibly because of imperfections in experimental

[1] Lecture at Hilversum, autumn 1948.
[2] *Soil Conditions and Plant Growth*, 1950, 25.
[3] Ibid.

technique, but the few recent accurate experiments that have been made have given no indication that the presence of farmyard manure or composts in a soil has any specific action in increasing the vitamin content or nutritive value of the plant.' He gives two references in support of this claim. Let us examine them.

One is a paper[1] summarizing the various factors influencing vitamin content. As far as the connection with the soil is concerned, only four vitamins –vitamin C, carotene, aneurin and thiamin—are studied, and none very thoroughly. Whereas, in fact, to-day some two dozen vitamins are known and others are constantly coming to light. So as far as vitamin content is concerned we have not much evidence there to support either view.

The second reference is more interesting. It concerns an experiment by the University of California[2] to find out the influence if any of organic matter in the soil on food fed to animals. Ten guinea pigs were used. Four were fed on Astoria Bent grass grown in a nutrient solution, replenished from time to time. The other six were fed on the same species of grass from a plot of soil. No other food was given. Let us first examine the manuring of the plot of soil. The grass was laid down in six plots, each of 60 sq. ft., i.e. about 6½ sq. yds., four years previous to the experiment. In the autumn, before sowing, 75 lb. of sheep manure and 25 lb. of alfalfa meal were added to each plot, while in the spring a further 30 lb. of farmyard manure and ¼ lb. each of sulphate of ammonia and calcium nitrate were added. Why such enormous doses of manure were necessary simply to grow grass is not stated, the losses of nitrogen must have been great. This treatment was repeated the following autumn and spring. Thereafter, for the next two years, they received a monthly dose of ¾ lb. of ammonium phosphate or ¾ lb. of calcium nitrate alternately. It should also be recorded that for the first two years each plot received ¾ lb. of lead arsenate—no doubt for wireworms or leatherjackets.

As shown earlier in this book, nitrogenous manures are destructive of soil organic matter, and it seems, from the need to control soil pests, which normally live on organic matter, that the large amounts of organic matter were being destroyed as fast as they were added. In any case, after a further two years of artificial manuring, it seems likely that there would be very little organic matter left in the soil apart from the plant's own roots and immediate remains. So, in fact, we have not so much a comparison between the influence of organic

<hr>

[1] Nutritional Abstracts 1943, 13, 155.
[2] D. I. Arnon, H. D. Simms and A. F. Morgan, *Soil Sci.*, 1947, No. 63.

manure versus nutrient solutions, as between soil with artificial manuring versus nutrient solutions. While not denying the existence of some organic matter in this soil, it is a very different matter from the exclusively organic manuring obtaining in Nature and on organic farms.

However, let us take a look at the guinea pigs. These, for a start, were raised on an artificial diet of whole-milk powder, whole-wheat meal, wheat germ, with salt and vitamins A and D added. Lettuce was given twice weekly. This diet for the first three weeks. Their ages at the start of the experiment were 4–6 weeks. In the intervening period they had Rabbit Pellets and wheat germ, with vitamins A and D from fish oil. Lettuce as before.

They were given 100 grammes of grass daily at first, but later they were given grass *ad lib.*, and often ate 300 grammes daily. The experiment continued for twelve weeks—a short period even for guinea pigs. No reproduction is recorded though sexes were mixed. All showed good growth in length, excellent skeletal and muscular development, with good fur and clear eyes and 'all other indications of nutritional well being'. They were, however, short of fat, presumably due to the low calorific value of the diet.

In spite of this, however, two of the grass-from-soil-fed animals died in the third week and one from the nutrient-solution grass in the sixth week, in spite of their apparent health. Compare this with Dr. McCarrison's thousand rats kept for over five years without any deaths from disease[1] or natural causes. Can it be said that their organic feeding, admittedly more varied, had no influence on their health? We need more experiments between organically grown food and artificially grown foods, designed as feeding tests. They are not difficult to do, and I am quite sure that if there really was no difference our fertilizer industry would have been only too keen to broadcast the details.

Apart from those more subtle deficiencies in foodstuffs there are the more direct and obvious results of mineral ill balance. 'Swayback' in sheep is due to a deficiency of copper in the herbage. It may be cured by feeding minute quantities of copper salts to the sheep or applying them to the soil. Here we have another case of the plants apparently in good health food yet not supplying the animal with its needs. Here are more examples.

Contagious[2] abortion and undulant fever in human beings has

[1] Sir R. McCarrison, *Nutrition and National Health*, p. 21.
[2] Allison, referred to in *Soil Ass. Journal*, Vol. 6, No. 1.

been well demonstrated as being associated with a deficiency of cobalt and manganese. Deficiencies of magnesium and phosphate are commonplace to-day, more common in fact in the stock of the 'good' farmer than his less-advanced neighbour. It is known, for instance, that the use of lime can restrict the uptake of phosphorus, and an excess of potash or lime that of magnesium. A shortage of calcium in the diet of cows leads to milk fever, but to lime a sour soil low in phosphates would render the phosphate even less available. By the use of organic matter and good husbandry, both phosphate and calcium can be released. Furthermore, it is of little use feeding calcium in an inorganic form to supplement lime-deficient fodder, as shown by experiments at the University of Missouri. On the other hand, if available phosphates are plentiful in the soil, calcium aids the assimilation of both phosphorus and nitrogen in the plant, giving a much better food product.

The mineral content of the herbage is often a reflection of the mineral availability of the soil, as shown in the following table.

PERCENTAGE OF PHOSPHORUS IN GRASSLAND HERBAGE
(DRY MATTER)[1]

	Good Soil		*Poor Soil*
Romney Marsh	0·26	Transvaal Veldt	0·04
A Scottish Pasture	0·32	Bechuanaland	0·02
		Victoria	0·04

This may be more closely connected in these cases with the climate than the total amount of phosphorus in the soil. The first two are from moist temperate climates with ample opportunity for microbial activity, while the last three are from hot dry areas—which are commonly overgrazed, restricting the plant root range, and organic content of the soil.

Animals can often detect mineral differences in their food—cattle and sheep show a tendency to graze the less-acid portions of their pastures.[2]

The accent on potash for tomatoes in glasshouses has made a magnesium deficiency in the plant commonplace, particularly on sandy soils, where both potash and magnesium are likely to be short. Though such heavy dressings of potash are not likely on farmland, magnesium deficiency would follow if the element was already low.

[1] J. W. and E. W. Russell, *Soil Conditions and Plant Growth*, 1950, 436.

[2] W. R. G. Atkins and E. W. Fenton, Sci. Proc. Roy. Dublin Soc., 1930, 19, 533.

A deficiency of magnesium is not an unusual complaint in cattle, especially if grazed on the quick-growing, reseeded popular leys used to-day.[1] It has been suggested, too, that the incidence of cancer is high in areas where a deficiency of magnesium exists in the soil.[2]

It must not be denied, however, that some natural soils are deficient in certain trace elements and would not produce anything at all but for the use of a few pounds to the acre of this deficient material. Take molybdenum and boron, for instance.

Whether such soils should be used for agriculture at all is a debatable point. They are normally to be found overlying the old primary rocks, which, being formed from the earth's molten mass, had a tendency to separate into layers and patches of different minerals. If, therefore, a primary rock is deficient in one mineral it may well be deficient in others also. By far the greater part of the earth's surface consists of sedimentary rocks, which, by their origin, are inevitably well mixed and likely to contain everything, more especially those formed under marine conditions.

These few examples are an indication of the complexity of the relationship of the various minerals to one another and of the difficulties of artificial manuring in relation to nutrition. One may do the right thing, but very often it is a case of 'out of the frying-pan into the fire'.

The use of heavy nitrogen dressings is not without its dangers. An article by Dr. J. K. Wilson, of the Department of Agronomy, Cornell University, New York,[3] shows the results of an analysis of nitrates in food. These are reduced by bacteria in the alimentary canal to nitrites which have a damaging effect on the haemoglobin of the blood. The use of nitrogenous top dressings to spring leys may be connected with the incidence of bloat in cattle for the same reason. From quantities found in leafy vegetables, frozen foods, and prepared baby foods, he suggests that it may be causing toxic and lethal symptoms in babies and adults. The high concentration of nitrates is considered to be due to the large applications of nitrate of soda to produce heavy yields of a bright green colour.

The cause of caries in teeth is a controversial question in the dental profession. That food influences their formation is undoubted—whether by its decay and subsequent dissolving action on the teeth

---

[1] *Farmer & Stock-Breeder*, 27th December 1949; Mr. C. M. Edwards, *Times Agric.* correspondent, 9th January 1950.
[2] E. Pfeiffer, Caxton Hall Lecture, 13th July 1950.
[3] *U.S. Agronomy Journal*, Jan. 1949.

or by its influence on tooth formation, is still a matter of much argument. There is little doubt in most minds that both play a part. An interesting sidelight on the influence of the soil is shown by the figures of American naval recruits: those from the drier central areas well supplied with minerals had only about two-thirds the caries of those from the leached and acid soils of the east.[1] It is often noticed, too, that their incidence is higher in districts using 'soft' water— more especially industrial areas, with their lack of sunshine and vitamin D.

The late Weston Price, a dentist,[1] spent a great deal of time studying primitive peoples in many parts of the world and found consistently that those races eating food as they found it, growing wild or on their cultivated plots, had consistently good teeth. As soon as they partook of civilized diets caries were the first defect to develop, and very quickly. He gives an example of the mineral content of the primitive Maori diet over that of his civilized counterpart. It contained 6·2 times as much calcium, 6·9 times as much phosphorus, 23·4 times as much magnesium, 58·3 times as much iron, and 10 times as much fat-soluble vitamins. The primitive people had 0·01 per cent of caries, while the civilized 55·3 per cent. Even the most vigorous refining of food would hardly account for these differences in mineral content.

It should be remembered in this connection, too, that our bodies and that of other animals are not the static organisms they appear at first sight. With the technique involving the use of radioactive isotopes it has been shown that the elements of which we are composed are being continually replaced, some more rapidly than others, but even our bones and teeth are subject to this replacement. It seems, then, that we need a steady and plentiful supply of these elements all through our lives.

The influence of fluorine in the diet on teeth has been made an excuse for medicating the drinking water in some places, and it is being advocated in this country. The possibility that such an addition will mean that those already receiving adequate amounts in their diet may then be receiving too much seems to have been overlooked. Safer by far to ensure adequate amounts by eating the unrefined foods grown in soils in which adequate organic matter enables such minor elements to be absorbed by the plant.

There have been a number of claims that farm animals prefer

---

[1] W. A. Albrecht, *Dental Journal*, Australia, Jan. 1951.
[2] Weston A. Price, *Nutrition and Physical Degeneration*, 1950.

naturally or organically grown fodder to that grown with artificials. Our research stations do not seem to have considered these claims deserving of investigation. Even the comparatively ill-developed sense of taste of the human being can often detect a difference. Certainly people tell me that my vegetables and fruit have a better flavour than those from the shop. Whether this is a matter of freshness or imagination is sometimes hard to say. The variety of vegetable or fruit, too, plays a part. But these factors are not always applicable.

Not much is known about the extent of the part played by bacteria in animal and human digestion. It seems that it is considerable in ruminant animals. Bacteria are often very particular as to their requirements in unexpected ways. When investigations were being made on the nitrogen-fixing bacteria at Rothamsted[1] it was found that glucose sterilized by heat was very toxic to those bacteria, but not when sterilized by filtration. My own digestion seems to confirm a similar effect. White sugar is difficult of digestion, while the brown varieties are not. Many people find a similar difficulty with boiled sweets and toffee, but not so much with chocolate. It may be that bacteria play a part in the human digestion of sugar, perhaps the impurities or reducing sugars present in unrefined cane sugar help to digest the whole.

Recent research has provided another complex story of soil bacteria and digestion.

A disease of sheep known as 'pining' occurs when the proportion of cobalt in the soil is below 5 parts per million, which suggests a need of 1 milligram daily. On the other hand, anaemia, which is one of the symptoms of pining, is cured by a daily dose of a natural cobalt compound of one-millionth of a gramme. This substance is variously known as vitamin B12, or the 'animal protein factor', or the 'pernicious anaemia factor'.

It has since been shown that bacteria in a sheep's stomach collect cobalt from the sheep's food. Perhaps it is the bacteria that need the cobalt—either to aid the sheep's digestion, or to manufacture vitamin B12, or some other necessary substance for the sheep.

This vitamin B12 is certainly manufactured by a soil actinomycete, *Streptomyces griseum*—the parent also of the antibiotic streptomycin.

The presence of this vitamin enables poultry and pigs to live on a vegetarian diet; and until its production, as a by-product of streptomycin, these animals needed a small proportion of animal protein.

[1] Dr. J. Meiklejohn, Rothamsted Report, 1950.

Hence its name of 'animal protein factor'. It seems likely, then, that completely vegetarian animals are provided with the vitamin by their intestinal flora.

Digressing for a moment on the needs of human beings for vitamins, the literal interpretation of our daily needs from laboratory evidence needs taking with caution. We are told that the normal daily vitamin requirement of the human body is so many units of vitamin A, so many of B, and so on.

These figures are obtained by experiments on human guinea-pigs, medical and biological student volunteers most commonly—healthy ones, of course. However, by far the greater bulk of the populace, as shown before, are unhealthy. These are, in fact, the normal people. (Or should one say 'average'?) These young students are those who are immune to, or have not yet begun to suffer from, the deficiencies of our national diet. Unwittingly, the experimenters have selected those guinea-pigs who can remain healthy on a bare minimum of vitamins. We are the ones whose requirements should be estimated, we, the great unhealthy. It seems likely that the medical world would find the results revealing. I know from personal experience that my own (and my wife's) need of fat-soluble vitamins is in the region of three times the 'normal', and of B vitamins almost double. Whether the need is an after effect of a lifetime of faulty feeding is immaterial to us.

It is reasonable to assume for evolutionary reasons that the ideal food for a young animal is that provided by its mother. It provides a balanced diet designed to build up the animal's digestive system until it is able to cope with the varied and more bulky food of the adult. (Natural evolutionary economy sees to it that the milk is put to the best possible use. We, in our wisdom, believe we can replace it.) It is also necessary in Nature for the animal to develop quickly so that the defenceless stage is passed in the shortest possible time. Though the latter reason no longer obtains for domesticated animals, it is replaced by the profitability factor. Profitability also works in another direction. The milk is a valuable cash commodity and provides immediate profit. Whether depriving the young of its milk is, in the long run, profitable is a contentious matter, research along these lines is always short-term and of little value. But there is no doubt that beef cattle, generally allowed to suckle when young, are much healthier and longer-lived than milking animals. It is estimated that the working life of the milch cow is two to three years, on the average, in Britain. Some are known to milk for twelve to fifteen years. A record of their upbringing would make an interesting piece of research.

The same principle applies to the human young. The principle of milk production—the feeding of high-protein foods—is well known to the farmer, but, oddly enough, it is seldom applied to the civilized human species. We are governed by the same natural laws. The particular value of mothers' milk is shown by the following figures, which refer to the Infant Welfare Centre of Chicago from 1924–9:

Deaths of infants—breast fed	0·15 per cent
artificially fed	8·40 per cent

For every breast-fed death there were fifty-six deaths amongst the artificially fed. While it is true that medical science to-day would prevent many of these deaths, the liability to infection remains—most of them were due to respiratory infections and to a lesser degree digestive troubles—all too common these days. Even cows' milk, it seems, was not comparable with that of its own species. How much less desirable is the remarkable range of cereal preparations offered to babies at ages of 4 months onward—a baby's digestive system does not even begin to digest starch until six months, the necessary enzymes are absent. One does not have to seek far for the digestive and so-called teething troubles of the average infant. The only time my own child suffered any digestive trouble was after a course of a dextrinized cereal supplement, at the rate of a teaspoon of the powder twice daily, at the age of 4 months. This trouble became gradually worse for about a week until the supplement was stopped. That was the first and last baby food ever bought. It was found far more desirable for all concerned to feed the mother. Mine is not an isolated experience.[1]

As to the value of a fresh, well-balanced diet there is no doubt, but though it is probably the most important factor in our life and happiness, our children are seldom educated to the fact. While our adult population is blinded by the power of advertisement. It is well to give a brief outline of the experiments of Sir Robert McCarrison which showed, to an extraordinary degree, the value of a fresh, unprocessed and balanced diet.

The experiments for convenience were done using rats, which have several advantages from the experimental point of view, including very similar food requirements to the human being. Two years in the life of a rat may be considered as equal to fifty years in a human

---

[1] Dr. L. J. Picton, *Thoughts on Feeding*, VI, VIII.

being. It is interesting to note that rats kept on a diet of fresh whole-meal flour cakes, with butter, sprouted peas, raw fresh vegetables *ad lib.*, milk, hard crusts for their teeth, and a little meat and bone, with water to drink, kept in perfect health. Over a period of five years there was no case of illness, no death from natural causes, no maternal mortality, and no infantile mortality.[1] This parallels the case of so many primitive human races.

These were the rats used for the experiments. Sir Robert McCarrison was impressed by the way that certain races in India, where he was Director of Research on Nutrition for some time, suffered from particular diseases. The only common factor seemed to be the diet, which varied from race to race.

He, therefore, fed three lots of rats on the same food as three races of Indians—the Sikhs, the Madrassi and the Travancore. They were kept for two years. The rats on the first diet suffered no loss or disease. Many on the other diets died and suffered from a great variety of diseases, especially respiratory and digestive troubles. It is further interesting to note that those fed on the Madrassi diet suffered to the extent of 11 per cent from peptic ulcer, and those on the Travancore diet 29 per cent. These percentages agreed well with their incidence amongst their human counterparts.

A similar experiment, using the diet of the average British workman, was made with twenty rats for 187 days, or about sixteen years in the life of man. The diet was as follows: white bread, margarine, sweet tea with a little milk (of which the rats seemed inordinately fond), boiled cabbage and potato, tinned meat and jam. It is true that the average diet is probably somewhat better now in Britain. The results on the rats were remarkable. They made poor growth, were rough-coated, irritable and savage, and finally began to eat one another. Six died of pneumonia and most suffered from digestive troubles. The diseases resulting from this diet—and even the tea-drinking habit, are remarkably analogous to those of the British people.

The argument developed in Chapter I may well be applied to the evolution of animals also. Animals as we know them, and also the human race, have evolved on a diet of plants (directly or indirectly) grown in an environment of decaying organic matter. There is no doubt that plants do absorb a good many materials from the soil, even though they do not use them themselves. (Take, for instance, selenium, some of the new systemic insecticides and the 'taint' from

[1] Sir Robert McCarrison, *Nutrition and National Health*, 1944, p. 21.

the use of benzene hexachloride insecticides, as examples of poisonous materials taken up.) The animal may well have become accustomed to using these materials—take Rowland's observation on vitamin B—even though the plant itself seems quite healthy without them.

What is indisputable is that many human communities distant from civilization, and farms which have turned to organic methods, have a degree of health considerably above the average. Often strikingly so. This, no doubt, is very largely due to the use of unfragmented and natural foods as much as the system of agriculture employed, and it is very difficult, if not impossible, to separate the influence of the unfragmented food from the organically grown food under normal conditions. Where it has been done—by Rowlands, McCarrison and Pfeiffer—the result has been unanimously in favour of the organically grown food. Such experiments have not yet been recorded by any advocates of orthodox methods, so that we have no direct contrary evidence.

It should be noted, also, that though our life-span is longer than that of many primitive races, with consequently more opportunity for disease, we suffer more acutely, and at a much earlier age, from the degenerative diseases—caries of the teeth, indigestion, rheumatism, nervous disorders, and so forth. Our defences against pathogenic organisms are, with the aid of modern medical science, enormous, but we come off a poor second with the degenerative diseases—some of which, indeed, are unknown amongst many 'uncivilized' peoples. These people have one thing in common—they do not know, or cannot afford, artificial manures. The only manures they have are their own crop and animal wastes—or else they lead a nomadic existence and live on naturally occurring foods.

To arrive at the perfect diet for an animal there are a number of approaches that can be made.

The animal's body may be analysed to get a knowledge of its constituents, and food of similar analysis, with additional energy-providing food calculated to provide the requisite amount of heat, is fed back to the animal. This, approximately, is the system chiefly in vogue to-day. It has severe limitations. We can only analyse an animal's body or food to a very limited degree. Perhaps twenty or thirty years ago this approach appeared to put the matter within our grasp. The result of further research, however, was to expose our ignorance. There is, in fact, no reasonable prospect that we shall reach our objective by this approach, the subject is far too complex.

A second approach is that of feeding by trial and error, a method frequently combined with the previous one and, though suffering from obvious limitations, has given quite good results. But as conditions vary the results cannot be consistent.

A third approach, largely neglected, is to study the diet of the animal in its pre-domesticated state. This, at least, will give us an ideal standard to measure by. The animal is obviously evolutionarily adapted to this food. It is true that they have, by artificial selection, been somewhat altered in their constitution—but this has been almost entirely a matter of shape and size—animal breeders almost automatically feed animals on the food they find most suited to them. It is a paying proposition to do pedigree animals as well as possible— a long life is an essential.

This evolutionary principle applies with much more force to human diet, where no artificial selection takes place and life is comparatively long.

It seems that white races have evolved on a diet consisting largely of animal foods—there being little else available during most of the year of a sufficiently energy-providing nature. Herbivorous animals of cold and temperate climates have thick layers of fat and hair for conservation of heat, and enormous abdominal capacities for extracting nourishment from the low-food-value vegetable matter that is to be found in such countries at most times of the year—but we have not.

Our food at present consists largely of vegetable starch, especially cereal starch. Cereals have come into popular usage in north Europe only during the past two thousand years, or perhaps a hundred generations, far too short a time for evolutionary adaptation to take place. It is only the extraordinary adaptability of the human organism that has permitted such a change—and such has been our increase upon it that we seem to be now bound to this diet. Such a change of food will cause a strain on our metabolism and may be another cause of our poor health as compared with primitive races.

It is an astonishing presumption on the part of civilized man that, with his limited knowledge, he should think he knows best how to feed plants and animals. He scarcely knows how to feed himself.

XI

ORGANIC AGRICULTURE
AND THE FUTURE

There is sufficient evidence available to-day to justify extensive research on organic farming methods, not merely to vindicate these methods but to improve on them. Several hundred farmers in Britain alone are obtaining yields above the national average, partly by good management, but the fact that they are not using the usual adjuncts considered necessary for good farming—artificial fertilizers, and deep cultivations—is worthy of investigation. The last two on the grounds of national economy alone, and, in addition, there is the evidence of better nutrition of crops, stock and human being. There is also a matter of increasing urgency—that of the disposal of urban wastes.

There are a number of reasons why this research has not taken place before.

Firstly, I think, is the fact that no clear exposition of the methods and principles involved has been set forth. It is hoped that this book will remedy the deficiency. While I make no claim for the accuracy of my theories, they are not unreasonable, and the facts presented are indisputable. We have had research on comparisons between organic and artificial manures, but these are irrelevant when their method of use is not considered: they are, in fact, no more than a red herring.

The second factor is allied to the first. It is the 'faith' of the compost writer and farmer. While faith may be considered a virtue in some connections, when associated with a search for knowledge it is any-thing but. It is a pernicious attitude of mind that saps a man's initiative and blinds his powers of observation and inquiry, an atti-tude of mind fostered in the past, and even now by those in high places, to ensure the subservience of those in their charge. This atti-

230

tude is not uncommon in the orthodox technician also, more especially at lower levels of responsibility. He is unwilling to cast aside long-held concepts when contrary evidence appears. This is a natural consequence of the fact that the scientist is nearly always right.

Having had some experience myself of the laboratory and of the field, I have come to realize the gulf between them. In the laboratory, controlled conditions, neat lists of causes and effects, all classified and pigeon-holed. In the garden, conditions are almost entirely out of our control with an endless variety of combinations and permutations of causes and effects. This is where the application of laboratory knowledge often comes unstuck. Laboratory techniques are used under conditions inapplicable to them. The technicians should approach their problems from the other end. Instead of pressing their findings on to the conditions obtaining in the field and then trying to find out why they don't work, study the plant under its natural conditions and find out why it does work.

A rigid laboratory discipline has been developed which demands that an experiment shall have only one variable, to be measured against a controlled background. This discipline fails when applied to natural environments—they are always changing, and attempts to render them static, if successful in any degree, are themselves an interference with the environment. This difficulty may be surmounted by using the statistical method. A larger number of experimental results are taken under different conditions which has the effect of evening out the changing environment. Extreme accuracy is no longer required, the large numbers are the safeguard against error. The factors, or combinations of factors, being studied are measured against this statistical background. By this means the ecological background of the experiment is undisturbed. It is in this field that new and profitable adventures are opened in the world of science.

This is not to say that it is not being done. It is,[1] but by far the greater amount of research takes the first approach.

We have, of course, an exactly analogous situation in the medical world. The cure of plant and animal difficulties is so much more profitable to those individuals and concerns practising it that prevention is almost overlooked. This is not a reflection on the integrity of such individuals, but an indictment of the economic system under which we live. In any case, the average general practitioner has far too much to do propping up our failing health to educate the public

---

[1] For instance, the Ecological Research Foundation, sponsored by the Soil Association, is being formulated for this very purpose.

into better ways of living—though there is little doubt that many could do much more in this direction.

This is where faith comes into the picture again. Faith in the professional man, whether his profession deals with the animal, vegetable or mineral. It is indeed difficult for the sceptic to contradict him with his limited knowledge. But when things go wrong he has a right to know why; and if this right was always exercised, the professional man would be compelled to question his own authorities, who in turn would be unable to take refuge in a welter of technical jargon.

A new approach is required by the composter also. He should not blind himself with faith and ethical considerations. Much needless opposition has been created to the organic cause by glib expressions of 'wholeness', 'natural cycles', and so forth. There is certainly much truth in them as applied to organic methods, but until the recipient has been conditioned to receive them it merely makes him suspect cranks at work. Because a thing is 'natural' it is not necessarily ideal from our point of view, and we must remember that even the compost heap, with its high temperature, does not occur in Nature. Agriculture itself is an interference with the natural order of things, but let us not blame modern agriculture for all its evils. The method of burning off the bush and forest for a short period of cultivation, as practised by so many primitive peoples, is more destructive of organic matter, and more conducive to erosion, than the use of artificial fertilizer. Indeed, the world's largest deserts were created by systems of agriculture used thousands of years ago, though possibly they were made more slowly.

Let us study things as they are, and not as we have made them. Let us question our beliefs to see whether they really fit the facts. If they don't, cast them out.

A third reason for lack of research is the lack of appreciation in Government circles for its need. When one considers the £1,640 million (Feb. 1954) spent annually on defence, the £1,500 million lost annually in sickness, it might be thought that the improvement of our first line of defence, not to mention our health from fresher and more wholesome food, would be deserving of more than a figure representing 0·3 per cent of agricultural production. The Government's hands are tied, of course; they must do what pleases the majority of the populace, or at least avoid displeasing them; their only hope is the gradual process of education—not an impossible one. Unfortunately, Britain being an industrial country, an interest in food will not be very great, unless it becomes short and the shortage seems likely to continue.

Research is influenced indirectly by a psychological aspect. Most people take the view that the more effort is put into a process or material the better it must be. Commercial interests have, not unreasonably, taken full advantage of the fact, pressed home by the power of advertisement. Whether it is a bigger and heavier tractor or the remarkable and unnecessary development of manufactured baby foods, people have come to appreciate these things, a demand has been created and research tends to follow. Not so much where it is needed but where it is most profitable.

Finally, research is largely limited by its capacity to measure. Thus the imponderable, quality, has been almost entirely ignored. Such aspects as have been studied are size, which is easily measured and often unrelated to other qualities, freedom from blemish by pest, disease or nutrition, and sometimes colour. Flavour and food value of crops have received practically no attention, oddly enough, though their food value is their purpose. It is not by any means impossible to measure food value by feeding experiments and it is not unlikely that this is connected with flavour. The only reason that can be deduced from this lack of research is that it has not been found to be profitable in a monetary sense.

It is appropriate here to expose one of the commonest refuges of the professional adviser and research worker, particularly in connection with organic methods. It is customary to say, for instance, 'no evidence has been found to show that organic methods are superior', or that 'the use of artificial manures has any harmful influence on the soil, or plant and animal health'. This is rather less than a half-truth. This claim implies that the evidence has been searched for. In fact it has not. It should be borne in mind by the protagonists of present-day methods that no experiments whatever of a comparative nature between organic and orthodox methods have been carried out by any of our national or commercial research establishments. The sole organization doing such work is the Haughley Experimental Research Farms Ltd., a private organization set up under difficult financial conditions for this very purpose. Our Government, at infinitesimal expense, could assist here in work of fundamental importance. The basic questions to which our national research bodies should find the answers are:

To what extent are fertilizers really necessary, and if they are, how can they be used to ensure no ill effects on soil, plant or animal?

Organic Agriculture and the Future

To what extent is it possible to farm economically, using only natural processes, harnessed and speeded to our advantage?

The answers will provide the country with real wealth.

Organic methods must stand or fall ultimately by their economic value. Either they pay the farmer or he will not use them. Ethics are poor food for the stomach or the bank manager. That these methods are economically feasible is borne out by the growing number who do practise them. Certainly many hundreds of farmers in this country, probably thousands by now, and an even larger number of market gardeners. But for those Doubting Thomases who think otherwise, there are a number of points that they should bear in mind. They will probably consider that the only difference is hauling 15 tons of dung against 5 cwt. of artificials. Or used to, until reading this book. The organic farmer need not haul much dung. With proper management, it can be dropped *in situ* with a considerable saving in nutrient value. Such as is dropped in the yards during winter is employing labour at a slack time, and composted manure is a much easier proposition to handle than dung. The capital outlay on a manure spreader is no more than that on a fertilizer drill. As regards the cost of the artificials (and lime to balance them), the farmer is merely employing another man in somebody else's factory for someone else's profit. It might well pay him better to employ an extra man or two, or pay his own men better. The artificials still require a man and tractor to spread them, though certainly the job is much quicker than spreading compost. Unfortunately, the artificials need spreading every year, and lime fairly often, while using the four-year ley and four-year arable rotation, compost need be spread only once every eight years, if that.

Further, the health of crops and stock on organically managed farms is incontrovertibly above the average. The average farmer hardly knows what diseases cost him. They cost the country £80 million in animal diseases alone—a tenth of the farmers' annual turnover. Vets' bills, medicines, crop sprays and machinery are obvious enough. The loss of production is incalculable, but very real. Even average crops are poor by comparison with the potential crops.

There is another advantage, intangible, but also very real—peace of mind. Will the corn lodge this year? Enough rain for the hay? (And fine weather?) The organic farmer can thrive under conditions that floor his neighbour. He takes advantage of Nature whenever possible, he doesn't work in spite of it.

So much for the individual aspect: what of the national outlook?

Organic Agriculture and the Future

In this world of increasing population and diminishing acreages, food prices are rising steadily—and will continue to do so into the foreseeable future. This is simply another way of saying food is getting shorter. It is in the national interest that as much labour as possible is turned to food production. This in turn will lessen the need for exports, releasing even more labour for the land. The fatuity of the 'export drive' will become apparent as labour returns to the land.

This state of affairs has come to pass because wages in factories are so much higher than on the land. This is a result of cheap food from overseas causing a depression of some eighty years in the farming industry. The situation is fast altering by the increasing scarcity of food and the increasing industrial power of food-producing countries. Agricultural labour is rapidly realizing its proper value in relation to the community. As world markets become saturated with our exports and the recipients find they have little food to offer us in exchange, the need for factory labour will diminish and will, by sheer force of circumstances, revert to the land. This country, in the course of the next twenty-five years, perhaps more, more likely much less, will become of necessity very much more of an agricultural nation.

This is likely to result in a fall in the standard of living as measured by the luxuries we enjoy to-day: there no longer being that easy money which sprang originally from the labour of the overseas food producers.

Happiness, however, does not spring primarily from the ownership of amenities and facilities, except to the kleptomaniac. Happiness springs from the sense of achievement; something made, something done; and there will be plenty of scope here.

Also, under the new conditions, people will be compelled to consume home-produced food in great quantity, being unable and unwilling to pay for the enfeebled foods offered to-day in cardboard packets, split into a variety of constituents, dyed and flavoured with coal tar. Health at least will be safeguarded.

Given an increased supply of labour on the land, our population can feed itself—but not on present methods. We must conserve our fertility, our plant food, probably even to the extent of returning a large part of our sewage wastes. Organic methods are essentially conservative. The soil is manured for the current crop and succeeding crops. It is not so with present methods. Plant nutrients are added every year far in excess of plant needs, with enormous consequential losses.

As an example of what is possible in the way of self-support, let us take Japan. Their standard of living is low by our ideals and, on account of their smaller size, they are able to live on about three-quarters of the food we eat. Whether their standard of happiness is any lower is another matter. Their standard of health is high when one considers the low incidence of degenerative diseases and infant mortality. They die rather younger than us, but when one considers that many of our old, and indeed some young as well, are kept alive by medical science, this is not an argument that the Japanese are less healthy. The fact that they live as long as they do under their more rigorous climate and conditions speaks volumes for their constitutions. The point to bear in mind is that, in spite of a similar population density to that of Britain, only one-sixth of their land is capable of cultivation, against rather more than half of ours. Yet they are largely self-supporting. We only produce half our food and precious little of our timber. Their food production, thus, is about four or five times that of ours per acre—we only need to double ours to be well fed.

Given a new approach we could live well on our own resources.

The international outlook is somewhat similar to our national one. It is considered by the Food and Agriculture Organization of the United Nations that world (September 1953) food supply needs to be 15–20 per cent greater in order that everyone should receive sufficient. In the major grain-producing countries there have been a succession of good harvests, with the result that the food supply is keeping pace with the increase in population though it has not yet caught up with it. This succession of good harvests has been due to better cultural methods, particularly that of leaving the soil unturned, but even more to plentiful rain during the growing season. A series of droughts, as occurred in the nineteen-thirties, is an ever-present possibility.

It may be added that the source of our fertilizers is not inexhaustible, neither are our present sources of motive power, though they may possibly be made good by other means. Our land too is limited, and, in fact, through lack of understanding, it is getting rather less. Mankind must learn in its own interest how to live within its means.

The human species has lived off the land for a very long time. Posterity will have to live on it.

GLOSSARY

AEROBIC: Taking place in the presence of air.

AMINO-ACIDS: 'Bricks' from which proteins are built.

AMOEBA: One-celled animal of jelly-like substance.

ANAEROBIC: Taking place in the absence of air.

ANTIBIOTIC: Substance excreted by an organism to poison its enemies or competitors.

ASEPTIC: In the absence of germs.

CARNIVOROUS: Flesh-eating.

CATALYST: Substance which facilitates a chemical reaction.

COLLOID: Jelly-like substance.

ECOLOGY: The study of the relationships of living things.

ENZYME: Similar to Catalyst—but secreted by living organisms.

LEACHING: The washing out of soluble substances by a solvent. Usually used in connection with soil and rain.

LEGUMES, LEGUMINOUS: Plants belonging to the clover or pea family. These have the power of utilizing atmospheric nitrogen, by the aid of certain bacteria.

LYSED: Dissolved.

METABOLISM: The life processes of an organism.

MICROFLORA: Microscopic plants.

MICRO-ORGANISM, MICROBE: Microscopic plant or animal.

MYCELIUM: The mass of fungal threads or tubes which ramify through the food material of a fungus. Taken individually, these threads are called HYPHAE.

MYCORRHIZA: 'Fungus roots.' The roots of a plant that are infected with a beneficent fungus.

NITROGEN: Literally, the element Nitrogen, which comprises four-fifths of the atmosphere. When used in connection with the soil, it is used to cover all compounds of this element in the soil—which are considerable.

NITROGENOUS MANURES: Those supplying compounds of nitrogen, suitable for assimilation, directly or indirectly, by the plant.

OSMOSIS: Certain materials, such as protoplasm, animal and plant membranes and living tissues generally, will permit small molecules or particles of substances to pass through them easily but large molecules with difficulty. Water, which consists of small molecules, will pass through easily; while sugar, for instance, consisting of large molecules, will not. If, then, we have a bag made of one of these 'semipermeable membranes', as they are called, containing a solution of sugar, and immerse it in water, the water will pass into the bag but the sugar will be unable to pass out. The result will be that the bag will commence to fill out and eventually considerable pressure will develop, depending on the strength of the sugar solution.

PATHOGENIC, PATHOLOGICAL: Causing disease.

PHOTOSYNTHESIS: The combination by plants, with the aid of light, of carbon dioxide and water to form sugar.

PHYSIOLOGICAL: To do with an organism's structure.

PRECIPITATED: Rendered insoluble.

PROTEINS: The components of Protoplasm.

PROTOPLASM: The jelly-like material contained by the cells of living organisms. This is the actual living substance of an organism and controls its life processes.

SAPROPHYTE: An organism living on decaying material.

SEMIPERMEABLE MEMBRANE: See OSMOSIS.

SYMBIOSIS: The living together, or partnership, of two (or more) distinct organisms for their mutual benefit.

TRANSPIRATION: The interchange of moisture and air between the leaf of a plant and the atmosphere, particularly the loss of water vapour from the leaf.

VIRUS: Particles too small to be studied by the optical microscope, which have the power of increasing in living organisms, causing disease.

INDEX

Acidity, soil, 31, 127, *et seq.*
  effect of compost on, 127
  tolerance of crops, 129
Actinomycetes, 50
Agriculture, development of, 57
  modern, 57
  organic, 57
Air in soil, 42, *et seq.*, 111
Albrecht, Prof. W. A., 13
Algae, 13, 14, 51
  blue green, 18, 52
Alkaline soils, 34
Aluminium and phosphates, 29, 105
Amino-acids in food, 64
  in soil, 28
Ammonia, 13, 90
Ammonium sulphate, 107
  phosphate, 107
Amoeba, 53
Analysis of plants, 120
  of soils, 118 *et seq.*
Anemones, 65, 163
Anthocorid bugs, 193
Antibiotics, 197
Ants, 33, 37, 208
Aphids, 66, 191 *et seq.*
Apple, 65, 157, 200
  quality exp., 165
Arable land, organic matter content, 95
Artificial manures, 60, 61, 63, 66, 67,
    69, 234
  and food value of crops, 63, 217, 228
Ash leaves, 30
Asparagus, 163
Aspartic acid, 16
Aspen, 30
Aspergillus, 198
Aspirins, 216
Atmosphere, minerals in, 28
Availability of minerals, 25 *et seq.*, 86
    et seq.
Azotobacter and plant roots, 19
  and algae, 52

Baby foods, 222, 226
Bacillus radicicola, 15 *ct seq.*
Bacteria, 50, 70
  in digestion, 224
  as humus formers, 72

Banana, Panama disease, 199
Barley, 79, 81, 83, 138, 140, 141
  minerals removed by 86, 122
  yields, 60, 61, 85, 86, 126
Beans, 30, 63, 141, 195, 196
  cultivation, 140
  minerals removed by, 157
  nitrogen fixed by, 17, 18
Bear, Dr. F. E., 77, 166
Bed System for market gardens, 162
Beet, 23, 24, 28, 30, 60, 64, 79, 81, 138,
    205
  cultivation, effect of, 117, 118
  sugar content, 64
Beetroot, 65
B.H.C., 210
  effect on plants, 213
Bio-dynamic agriculture, 185, 196
Biological activity and fertility, 15
Birds, 194
Biskind, Dr. M. S., 210
Blackcurrants, 65, 74, 193
Black nightshade, 138
Bloat, 63
Boron, 34, 222
Botrytis, 195
Boussingault, 21, 104
Broccoli, 66, 207
Broom, 30
Brown footrot, 198, 207
Buckwheat, 31
Bulbs, 163
Burnet, 98
Bushes, 15

Cabbage, 66, 74, 156, 160, 190, 209
Calcium, 31, 34, 105, 221
  cyanamide, 79
  sulphate, 106 *et seq.*, and whiptall,
    109
Cancer, 222
Carbohydrates, kinds of, 70
  manufactured by plants, 98
  and nitrogen fixation, 15
  in plants, 64
  in soil, 89
Carbon dioxide and photosynthesis, 13
  in soil, 30, 42
Carbon content of soils, 95

Carbon: nitrogen ratio, 89 *et seq.*
Carbon-nitrogen compounds in soil, 89, 95, 102
Caries, 222, 228
Carnivorous plants, 27
Carrots, 65, 100, 157, 161, 218
Carrot fly, 65
Caterpillars, 194
Cattle, 60, 222
Cauliflower, 34, 109
Cellulose, 13, 32, 70
Cereals and artificial manuring, 124
  in human diet, 229
Changing to organic methods, 135
Chase, Mr. J. L. H., no digging plots, 149
Chicory, 98
Chlorine, 34
Chocolate Spot, 63, 196
Chrysanthemums, 67, 69, 159
Citric acid, 29, 119
Cladosporium, 215
Clay, 36, 43
  effect of lime on, 127
  subsoil and worms expt., 32
Claypits, 15
Cleanliness of produce, 167
Cleveringa, Dr., 111, 218
Cloches, 65
Clostridia, 18
Clover, 58, 85, 115 (*see also* White Clover)
  in acid soil, 127
  clover sickness, 205
Clubroot, 208, 209
Cobalt, 221, 224
Coconuts, 208
Cocksfoot, 135, 168
Codlin moth, 208
Coltsfoot, 145
Combine straw, 189
Compost, 59, 63, 69, 78, 84, 172 *et seq.*
  air supply, 172
  analysis, 181
  application rates, 138, 160
  bacterial activity, 176
  biodynamic activators, 185
  clay in, 177
  costs, 181
  disease control, 185 *et seq.*
  dung, comparison, 186
  effect on soil, 91
  food of organisms in, 176
  insulation of, 173
  made with tomato haulms, 67
  materials, 173, 181
  methods, 177
  moisture, 174, 179
  and nitrogen fixation, 174
  and nitrogen needed, 174
  and nitrogen losses, 173, 180

  and nitrogenous manures, use with, 93
  and nitrogen, use of synthetic, 180
  Rothamsted, made at, 79, 84
  Rothamsted exp. with, 78 84
  sewage, use in, 182 *et seq.*
  sheet, 174, 183
  turning, 179
  use of, 183, 186
Conifers, 30
Contagious abortion, 220
Copper, 30, 34, 35, 214, 220
  oil emulsion spray, 66, 214
Couch Grass, 143, 145
Coward, Mr. R., 141
Crop Yields, 58 *et seq.*
Crowther, Dr., 84

D.D.T., 210 *et seq.*
  in food, 211
  in soil, 211
Decomposition of organic matter, 90
Deep-rooted plants, 98 *et seq.*
  and waterlogging, 129
Defence expenditure, 232
Degenerative diseases, 228
Denitrifying organisms 73 *et seq.*, 111
Derris, 66, 210
Diatoms, 51
Diet of animals, 228
Digestion and bacteria, 224
  of babies, 226
Disc harrow, 49
  and weeds, 142
Drainage of natural soil, 43
  of turf, 44
Drought, 116; and mildew, 195
Dry rot, 40
Dung, 67, 74, 120, 138, 172, 205, 234
  v. compost, 59
  in compost, 174
  hormones in, 101
  losses of nutrients from, 189
  need for, 101
  and organic content of soil, 91, 96, 113, 114

Earthworms—*see* Worms
Earliness of crops, 167
Ecological Research Foundation, 231
Eelworms, 53, 200 *et seq.*
  potato, 201
  predators, 200
  Rothamsted exp., 201
  in S. Rhodesia, 201
Elements in plants, 33
Enzymes on roots, 27
Erosion, 52, 130 *et seq.*
  run off under various crops, 130
Euglena, 53
Evolution, 13, 14, 227

Index

Eye Spot, 197

Faith, 230, 232
Fallow, 46, 114, 196
F. A. O., 30, 236
Farmland area, 29
Farm workers, crops produced by, 126
Fat hen, 138
Faulkner, E. H., 152
Ferns, 14
Fertilizers, cost of, 63, 234
    and continuous cereals, 123
    Fertility, effect on, 20, 112
    Food value, effect on, 63, 217, 228
    Micro-organisms, effect on, 108
    Worms, effect on, 109
Fertility of soil, 15, 94
    definitions, 102
    diagrammatic representation of, 94
    effect of manures on, 112
Field capacity, 39 et seq.
Fixation of minerals, 27, 104 et seq.
Flax, 194
Flour, 227
Flowers, 163
Fluorine, 223
Food, of animals, 227
    young, 225
    animal preferences, 221
    of British workman, 227
    world shortage of, 235
Food value of crops, 63 et seq., 217
Foot and mouth, 218
Formalin, 126
Freezing, effect on compost, 187
Frozen foods, 222
Fungi, 21, 50, 70
    as humus formers, 72
Fungicides, 66, 209, 213
Fusarium culmorum, 198

Galls on legume roots, 15
Gas from manure, 188
Gerretsen, F. C., 106
Glasshouse soils, 120, 164
Glucose, sterilized, 224
Granite, weathering of, 15
Grass, 15, 44, 124, 129
    in orchards, 115, 168
    and phosphate deficiency, 154
    and soil structure, 43, 44
Green manures, 74 et seq.
    bad effects of, 74, 91, 114
    as mature crops, 74, 114
Greenflies, 191 et seq., 213
Greensand, 61
Greenwell, Sir Bernard, 65
Grossbard, E., 197
Guinea pigs, feeding exp., 219

Haemoglobin, 222

Haughley Research Farms expts., 86 et
    seq., 109
    clover, 58
    silage, 64
    wheat, 58
Hay, 61, 150 et seq., 157
Health, 216 et seq., 235
    of Britain, 216
Heat in soil, 41 et seq.
    effect of cultivation on, 42
Heather, 34
Heathland, 107
Hedgebanks, 99, 111
H.E.T.P., 210
Hoeing, 117 et seq.
Honeydew, 192
Hoof and Horn meal, 66, 67, 181
Horses, 60
Howard, Sir Albert, 24, 82, 84, 151,
    169, 178, 193
Human diet, 229
Humus, definition, 35
    formation, 54, 89 et seq.
    losses by cultivation, 46 et seq., 111
    as nitrogen and mineral source, 19
    et seq., 35
Hydroponics, 111

Implements for organic methods, 146
Indigestion, 228
Ingham, M. C., 28
Insects, in soil, 53
Insecticides, 209 et seq.
Iron, 26, 34
    and Phosphates, 29, 105
Irrigation, 99 et seq.
Isotopes, 131
Italian Clover, 138

Japan, 236

Kale, 63, 138, 190
Krakatoa, 14

Laboratory, 231
Lacewing flies, 193
Ladybirds, 193
Lamb's-tongue, 138
Leach, R., 208
Leaching of nitrogen, 46
    losses in a compost, 79
Leather jackets, 193
Leaves, effect on soil, 30
    absorption of nutrients, by 28
Legumes, 15 et seq., 35
    and grass, 16, 17
    mowing, 115, 150
Lentils, 18
Ley, effect on cropping, 134
    deep-rooted (recipes), 135, 136
    treatment of, 136, 149

Light, 13, 70
Lignin, 13, 70
Lime, 52, 95, 118, 127 *et seq.*
  deficiencies induced by, 127, 221
  effect on cropping and soil organisms, 184
Limiting factor in agriculture, 133
Lincolnshire potato land, 111
Liss, Mr. G. O., 67
Lucerne, 18, 30, 98, 156, 161, 168, 170
  Cutting exp., 150
Lupins, 30, 98

Magnesium, 30, 31, 34, 105, 222, 223
Maize, 48
Malic and Malonic acids, 29
Manganese, 34, 221
Mangolds, 138, 157
Maori, diet of, 223
Market garden crops, 65
  machinery and cultivations, 162
  manuring, 156, 190
  soils, 64
  woburn exp., 85
Marrows, 65
Martin Leake, Dr., 41
Massey, Lt.-Gen. H. R. S., 60
Mayall, Mr. S., 60
MacCarrison, Sir Robert, 216, 220, 226
McDonagh, Dr., 131
Mealie bug, 213
Meat, 157
Medical science, 228, 231, 236
Medick, 161
Meta, 194
Microbes, 14
Micro-organisms, 50 *et seq.*, 70, 104
Mildew, 63, 195 *et seq.*
  on lettuce, 195
  on cereals, 196
Milk, 60
  butterfat, 60, 61
  mother's, 225 *et seq.*
  quality, 65
Millipedes, 53
Minerals, extraction by plants, 25 *et seq.*, 36 154
  lost by natural agencies, 104
  sources for market gardens, 158
Mites, 53
Molybdenum, 34, 35, 109, 222
Monoculture, 100 *et seq.*
Mosses, 14
Moulds, 195 *et seq.*
Mountain grazing, 154
Mowing of orchards, 115, 168 *et seq.*
Mulching, 44, 169
Mushrooms, 51
Mustard, 114, 138, 162

Mycorrhiza, 21 *et seq.*
  and artificial manures, 23
  needs of, 21, 23
  and plants, 21 *et seq.*, 107
  Rothamsted, work on, 24

Nematodes, 53
Nervous diseases, 228
Neville, Sir J. E. H., 58, 60
New Jersey, U.S., 106
Nickel, 34
Nicotine, 210, 215
Nitrate, in food, 222
  of soda, 222
  in soil, 90, 97, 193
'Nitro Chalk', 59
Nitrogen-carbohydrate ratio in plants, 16, 23, 192
Nitrogen, and cell walls, 196
  immobilization, 90, 188
  losses—by cultivation, 47 *et seq.*
  losses by leaching, 46
  losses in compost heaps, 173
  and mildews, 196
  in orchards, 169
  quantity in soil, 19, 89 *et seq.*
  in plant, 21, 192
Nitrogen fixation by legumes, 15 *et seq.*
  quantity fixed, 16, 17, 18
  left in soil, 16, 17, 18
Nitrogen fixation by free living organisms, 18 *et seq.*, 174
  needs of, 18
  and plant roots, 19
  quantity fixed, 18
Nitrogenous manures, 59, 72 *et seq.*
  and clover, 115
  on fenland, 47
  on grass, 152
  effect of heavy dressing on organic matter, 77
  ill effects of, 73 *et seq.*, 91, 110 *et seq.*, 131
  losses of, 73
Nitrification of humus, 21
Nitrifying bacteria, 20
Nutrients, in plants, 14
  solution, feeding exp. on animals, 219
  in worm casts, 31

Oakwoods, 194
Oats, 30, 60, 61, 126, 140, 150
Onions, 66, 157
Orchard management, 168
Orchids, 15
Organic *v.* Artificial manures, 230
Organic matter, 29
  effect on availability of minerals, 81, 84, 85, 89
  building up of, 133 *et seq.*

Index

Organic matter—*continued*
conservation of, 133 *et seq.*
cultivation losses by, 47 *et seq.*
decomposition, factors affecting, 96
and fertility, quantity needed, 98
and lime, 127
quantity in average soils, 113
quantity in various soils, 95
Osmosis, 25, 195
Oxalic acid, 26

Parasitic organisms, 13, 199
control of pathogens by, 208 *et seq.*
Parathion, 210
Parsnips, 66
Pasture, 95, 189 (*see* grass)
Pathogenic organisms, 124
in compost, 185
control by predators, 199, 208 *et seq.*
and sterilization, 124, 164
Peach leaf curl, 200
Peas, 17, 18, 59, 60, 61
Peat, 34, 35
Pedigree animals, 229
Penicillium, 51, 198
Peptic ulcer, 227
Persicaria, 139
Pests, resistance to insecticides, 215
Pewsey, Vale of, crops produced, 126
Pfeiffer, Dr. E., 59, 61
Phosphates, 26, 28, 34, 98, 105, 106, 107
availability, 29, 105, 122
in barley crops, 89, 122, 123
and clay, 26
in herbage, 154, 221
in organic matter, 123
in sewage, 29
in soil, 30
in soil solution, 26
in worm casts, 31, 32
Phosphate rock, 106
Photosynthesis, 13
Phytin, 27
Pigs, 60, 217
Pine Trees, 30
Pine Forests, 34
Pizer, Mr. D. H., 149
Plants, analysis of, 120
under artificial conditions, 131
cycle of growth and decay, 71, 132
evolution and soil, 45
growth in nutrient solution, 27
needs of, 15
Planting distances, 163
Ploughing, depth, 117
effect on decomposition of residues, 72 *et seq.*, 116
panning, 116 *et seq.*
reason for, 49
wet soils, 148

Poliomyelitis, 218
Potash, 23, 25, 28, 29, 30, 31, 34, 67, 80, 81, 105, 197, 205, 221
available, 106, 119
in bacteria, 164
salts, 158
Potato, 25, 78, 118, 119, 157, 165
blight, 199, 214
cultivation 66, 139
yields, 59, 60, 61, 66, 126, 160
Potting Composts, 159
Poultry, 178
Prairie soil, 18, 37, 95
Price, Dr. Weston, 223
Primary Rocks, 222
Protein, in plants, 63
Protozoa, 53
Pyrethrum, 210

Quarries, 14
Quality of produce, 165 *et seq.*
Quassia, 210

Radiobacter and rape, 24
Radish, 205
Railway cuttings, 14
Rape, 24
Rayner, Dr. M. C., 107
Red Shank, 139
Red Spider, 191, 213
Research, 230 *et seq.*
Rheumatism, 228
Rhizobium, 15
Rhododendrons, 34
Ringspot, 207
Rocks, 222
breakdown of, 13, 14
Roots, and bacteria, 15, 24, 26, 27, 197
development of, 14, 36, 116
depth of, 37
excretion by, 28
extent, 45
and mineral absorption, 25 *et seq.*
and large molecules, 25
taproots, 37
Rotary cultivation, 49, 143 *et seq.*
comparison with other methods, 143 *et seq.*
cultivators, 162
and soil structure, 145
Rotations of crops, 134, 140 *et seq.*
Rothamsted, straw and compost exp., 78 *et seq.*
soil, fertility of, 85, 96
Rowlands, M. J., 217
Runner beans, 66
Russell, Dr. E. W., 31, 218
Rusts, 195
Rye Grass, 114, 136, 168, 209

Salt, amount deposited by atmosphere, 28
   concentration in glasshouse soils, 69, 120
Salter, Dr. R. M., 153
Sand, 38, 43
Sainfoin, 98
Sawdust, 107, 173
Scab, 200
Schradan, 210
Scots pine, 30
Scott-Watson, Sir. J., 110
Sedimentary rocks, 222
Seed bed, 49
Seed sowing rates for cereals (*see also* Sowing), 62, 137
Sickness, cost of, 232
Silica, 196
Silicon, 34, 35
Silt, 37, 38, 43
Silverweed, 145
Shallow cultivation, 144 *et seq.*
Sheep, 83, 154, 221, 224
Sheet composting, 83, 188
   v. compost, 114
Shelter, 170
Slugs, 193
Smith, Dr. A. M., 110
Sodium, 34, 105
Soil, analysis, 118 *et seq.*
  animals, 53
    food of, 55, 56
    and weight of, 54, 55
  artificial structure, 111
  chalk soil, 60
  development of, 13
  fertile 50, 93 *et seq.*
  heat, 41
  kinds of, 37
  micro-organisms, food of, 89
  natural, 36 *et seq.*, 45, 71
  proteins, in 28
  puddling and worms, 55
  waterholding capacity, 38 *et seq.*
Sowing, 137 *et seq.*, 145
Soya beans, 18
Spanish moss, 29
Speedwell, 193
Spiders, 193
Sprowston, exp., 82
Starch, 70, 194
  in human diet, 229
Steaming of soil, 69, 124, 164
  coal used, 69
Sterilization of soil, 124 *et seq.*
  ease of re-infection, 164
Sterility, 61
Stomata, 195
Storey, Dr. I. F., 110
Straw, as manure or compost, 78 *et seq.*, 83, 107

sheet composting, 189
  fungi on, 197
Strawberries, 66, 160, 163, 204
Streptomyces, 198, 199
Subsoiling, 155
Sugar, 37, 70
Sulphur, 34, 213
  sulphate of ammonia, 109, 205
Sulphates, 107 *et seq.*
  and plant roots, 109
Sulphides in soil, 107, 148
Superphosphate, 105, 107
Surface cultivation and erosion, 186
Swayback, 220
Sweet clover, 18, 30, 98, 119, 170
  cultivation and uses, 153, 160 *et seq.*
Sykes, Mr. Friend, 60, 65, 121, 181
Symbiotic organisms, 13 (*see also* Rhizobium and Mycorrhiza)

Take all, 197, 206, 208
Tares, 114
Tartaric acid, 29
Temperate climates, 71
T.E.P.P., 210
Thistles, 143
Thrips, 208
Tilth, 111, 137, 145
Timothy grass, 168
Tomato, 66, 157, 160, 163
  composted haulms, 186, 209
  cost of growing, 68
  and mycorrhiza, 23
Trace elements, 26, 158
Transpiration, 28
Trees, 15, 133
Trefoil, 138, 161, 206
Trichoderma, 198
Tropical climates, 71
Turner, Mr. F. N., 60, 121
Turnips, 66, 126, 138, 157

Undulant fever, 220
Urine, 152
Utilization of natural elements, 132 *et seq.*

Vegetables, 65, 222
Vetches, 18
Virus, 27, 304 *et seq.*
  and aphides, 205
  inhibiting substances, 203
  and manuring, 204, 205
  in strawberries, 204
Violet root rot, 207
Vitamins, aneurin, 219
  B, 217, 218
  B12, 224
  C, 219

Index

Vitamens, aneurin—*continued*
  carotene, 219
  thiamin, 219

Wareham Heath, 107
Waste Hops, 107
Water, conservation, 38
  evaporation from soil, 41
  and fungal hyphae, 40
  for germination, 39
  movement in soil, 39, 115
  in plant, 37
  reserves in soils, 37, 116
Waterlogging, 34, 43, 129
  and diseases, 207
  natural vegetation on soils, 148
  treatment of soils, 148
Weather, effect on plant food supply, 72
Weeds, 142 *et seq.*
  in compost, 172
  destruction, 137, 142 *et seq.*
  germination and temperature, 138
  in market gardens, 163
  perennial, 147
  effect on tilth, 137, 147
  uses of, 146
Weed killers, 147 *et seq.*
Weiss, A. P., 32

Wheat, 47, 48, 85, 157, 134
  rotations, 140
  sowing of, 136 *et seq.*
  yields, 58, 59, 60, 61, 126, 134
Whiptail, 34, 109
White clover, 16, 33, 168 (*see* Clover)
White fly, 67, 208
White grub of sugar cane, 208
White sugar, 224
Wild oats, 142
Wilson, Dr. J. K., 222
Wind, 170
Winter flooding of glasshouses, 68
Wireworms, 193, 194
Woburn, winter barley expt., 83
  Market garden manures expt., 85
  soil acidity expt., 108
Woodlice, 53
Woolly aphis, 208
Worms, 54 *et seq.*, 71
  availability of minerals in casts, 31 *et seq.*
  casts, 56
  and fertilizers, 109
  holes, 37
  soil moved by, 56
Wylie, Mr. J. C., 182

Zinc, 34